BIOMARKERS OF POSTPARTUM PSYCHIATRIC DISORDERS

BIOMARKERS OF POSTPARTUM PSYCHIATRIC DISORDERS

Edited by

JENNIFER L. PAYNE
Johns Hopkins University
Baltimore, MD, United States

LAUREN M. OSBORNE
Johns Hopkins University
Baltimore, MD, United States

ELSEVIER

ACADEMIC PRESS
An imprint of Elsevier

Academic Press is an imprint of Elsevier
125 London Wall, London EC2Y 5AS, United Kingdom
525 B Street, Suite 1650, San Diego, CA 92101, United States
50 Hampshire Street, 5th Floor, Cambridge, MA 02139, United States
The Boulevard, Langford Lane, Kidlington, Oxford OX5 1GB, United Kingdom

Notices
Knowledge and best practice in this field are constantly changing. As new research and experience broaden our understanding, changes in research methods, professional practices, or medical treatment may become necessary.

Practitioners and researchers must always rely on their own experience and knowledge in evaluating and using any information, methods, compounds, or experiments described herein. In using such information or methods they should be mindful of their own safety and the safety of others, including parties for whom they have a professional responsibility.

To the fullest extent of the law, neither the Publisher nor the authors, contributors, or editors, assume any liability for any injury and/or damage to persons or property as a matter of products liability, negligence or otherwise, or from any use or operation of any methods, products, instructions, or ideas contained in the material herein.

Library of Congress Cataloging-in-Publication Data
A catalog record for this book is available from the Library of Congress

British Library Cataloguing-in-Publication Data
A catalogue record for this book is available from the British Library

ISBN: 978-0-12-815508-0

Publisher: Nikki Levy
Acquisition Editor: Joslyn Chaiprasert-Paguio
Editorial Project Manager: Timothy Bennett
Production Project Manager: Sruthi Satheesh
Cover Designer: Matthew Limbert

Typeset by TNQ Technologies

Contents

Contributors

Jonathan S. Abramowitz
Department of Psychology and Neuroscience, University of North Carolina at Chapel Hill, Chapel Hill, NC, United States of America

Jennifer L. Buchholz
Department of Psychology and Neuroscience, University of North Carolina at Chapel Hill, Chapel Hill, NC, United States of America

Kristina M. Deligiannidis
Women's Behavioral Health, Department of Psychiatry, Zucker Hillside Hospital, Northwell Health, Glen Oaks, NY, United States of America;
Departments of Psychiatry and Obstetrics & Gynecology, Zucker School of Medicine at Hofstra/Northwell, Hempstead, NY, United States of America;
Feinstein Institute for Medical Research, Manhasset, NY, United States of America

Arianna Di Florio
Cardiff University, Cardiff, United Kingdom;
University of North Carolina at Chapel Hill, Chapel Hill, NC, United States of America

Christy Duan
Department of Psychiatry, Zucker Hillside Hospital, Northwell Health, Glen Oaks, NY, United States of America

Jerry Guintivano
Department of Psychiatry, University of North Carolina at Chapel Hill, NC, United States of America

Liisa Hantsoo
Department of Psychiatry, The University of Pennsylvania Perelman School of Medicine, Philadelphia, PA, United States of America

Megan Hare
Department of Psychology, Florida International University, Miami, FL, United States of America

Samantha N. Hellberg
Department of Psychology and Neuroscience, University of North Carolina at Chapel Hill, Chapel Hill, NC, United States of America

Zachary Kaminsky
DIFD Mach-Gaensslen Chair of Suicide Prevention Research, The Royal's Institute of Mental Health Research (IMHR), Associate Professor of Cellular and Molecular Medicine, University of Ottawa, Ottawa, Ontario, Canada

Mary Kimmel
Department of Psychiatry, University of North Carolina School of Medicine, Chapel Hill, NC, United States of America

Jamie Maguire
Tufts University School of Medicine, Department of Neuroscience, Boston, MA, United States of America

Katherine McEvoy
Perinatal Psychiatry, The Johns Hopkins University School of Medicine, Baltimore, Maryland, United States of America

Samantha Meltzer-Brody
Department of Psychiatry, University of North Carolina at Chapel Hill, NC, United States of America

Gerald Nestadt
Department of Psychiatry and Behavioral Sciences at Johns Hopkins Medical Institutions, Baltimore, Maryland, United States of America

Lauren M. Osborne
Department of Psychiatry & Behavioral Sciences, The Johns Hopkins University School of Medicine, Baltimore, MD, United States of America;
Department of Gynecology & Obstetrics, The Johns Hopkins University School of Medicine, Baltimore, MD, United States of America;
Women's Mood Disorders Center, The Johns Hopkins University School of Medicine, Baltimore, MD, United States of America

Jennifer L. Payne
Department of Psychiatry & Behavioral Sciences, The Johns Hopkins University School of Medicine, Baltimore, MD, United States of America;
Department of Gynecology & Obstetrics, The Johns Hopkins University School of Medicine, Baltimore, MD, United States of America;
Women's Mood Disorders Center, The Johns Hopkins University School of Medicine, Baltimore, MD, United States of America

Karen T. Putnam
Department of Psychiatry, University of North Carolina at Chapel Hill, NC, United States of America

Jack Samuels
Department of Psychiatry and Behavioral Sciences at Johns Hopkins Medical Institutions, Baltimore, Maryland, United States of America

Lindsay R. Standeven
Department of Psychiatry and Behavioral Sciences at Johns Hopkins Medical Institutions, Baltimore, Maryland, United States of America

Patrick F. Sullivan
Department of Psychiatry, University of North Carolina at Chapel Hill, NC, United States of America;
Department of Genetics, University of North Carolina at Chapel Hill, NC, United States of America;
Department of Medical Epidemiology and Biostatistics, Karolinska Institutet, Stockholm, Sweden

Preface

For the last several years, the academic medical world has been abuzz about "personalized medicine"—the tailoring of medical treatment to characteristics that, if not unique, are specific to individuals. This approach assumes that many medical disorders may carry a single name but in reality be heterogeneous disorders, with many pathways toward a common end. To create truly personalized medicine, we need to have a full understanding of the varied processes—from genes, to gene expression, to environmental influences, to changes at the molecular and cellular level—that go into creating the symptoms of a disorder in any particular individual.

One hallmark of that understanding is the search for biomarkers, "quantifiable biological parameters" (Selleck, Senthil, & Wall, 2017) that can indicate the presence of disease or response to treatment. Some of the earliest biomarkers in medicine were noticed nearly 200 years ago, though with little understanding of their actual function, and the NIH now defines many types of biomarkers across most fields of medicine (FDA-NIH Biomarker Working Group, 2016). Biomarkers can serve many functions—they can help to diagnose a condition (as with hemoglobin A1c for type 2 diabetes), or to monitor a known condition (as with prostate-specific antigen [PSA] for those with prostate cancer). They can measure the response to treatment (for example, blood pressure for patients with hypertension given an antihypertensive treatment) or predict the likelihood of response to an exposure (for example, BRCA1 and 2 to evaluate which women will respond to certain ovarian cancer treatments). They can also be prognostic—identifying the likelihood of a clinical event or recurrence (BRCA1 and 2 to predict who will develop a second breast cancer) or the likelihood of disease developing among those currently without disease (again, BRCA1 and 2).

The idea of the biomarker of risk is enormously appealing—a "liquid biopsy" (Selleck et al., 2017) would allow those with a high likelihood of disease to get serial monitoring that is noninvasive, simple, and far less expensive than many monitoring techniques (such as imaging). The reality, though, is that in many fields biomarkers are still in the world of research and have not yet made it to a clinically useful stage. To be clinically useful, a biomarker must give us more information than whatever we currently use to predict risk and must be specific to the disease in question. It is not useful to find a protein that is elevated in people who will go on to develop a particular disease if that protein is also elevated in those who will *not* develop the disease. Nor is it useful to find a biological means to predict disease if it does not increase our prediction ability above means we already have, or by means of a test so expensive that it cannot be used by the majority of people. When we apply these criteria, we realize that while many purported biomarkers have been discovered through exploratory scientific approaches, few have been replicated in diverse populations and developed into clinically meaningful tests.

Nowhere is this more true than in psychiatry. Other fields of medicine have at least a smattering of somewhat useful biomarkers—again, think BRCA1 and 2 in cancer—but the search in psychiatry has proved elusive. The search itself has been energizing, with researchers employing many fields and techniques (genetics, epigenetics, neuroimaging, electrophysiology, the inflammatory system, and others). Moreover, the appeal of a biomarker in psychiatry is perhaps greater than in other fields of medicine, for we have no uniform objective means to diagnose psychiatric illness. Our current methods, which rely on the Diagnostic and Statistical Manual to categorize related groups of symptoms into heterogeneous diagnostic categories, are insufficient, and the allure of a specific biological test is mighty—yet the very nature of our diseases, and the complex roots of their causes blending biology with environment, has made this search elusive. There are no currently accepted specific biomarkers in psychiatry (Lozupone et al., 2019).

So if there are no biomarkers in psychiatry, why this book? This book exists because we view the perinatal period as potentially the most useful timepoint in the life course to search for these elusive biomarkers, and we wanted a chance to collect here the current state of that search. The two of us came at our interest in perinatal mental illness, and our conviction of the unique usefulness of this period for scientific discovery, from two very different perspectives. One of us (JLP) came from a neuroscience background, and spent her early training years wondering what the "broken part" might be in mood disorders, including major depression and bipolar disorder. By that time, it had become clear that research was not going to identify a single genetic or biological etiology for either major depression or bipolar disorder, that these illnesses were likely the result of polygenetic vulnerabilities plus environmental influences, and further, that these illnesses were likely to have multiple underlying biological "broken parts." As she thought further about how to try to identify more homogenous subgroups that might allow us to study the underlying biology, she suddenly realized that postpartum psychiatric illnesses, which certainly appeared to be triggered by the massive hormonal changes that accompany delivery, might be one such group. The predictable timing of the postpartum time period would allow us to study a group with a known timeframe for developing disease, and the insights thus gleaned could then be applied back to other psychiatric illnesses. The other of us (LMO) came from an obstetrics and gynecology background, looking at perinatal mental illness as a morbidity of pregnancy. She realized early in her training that there are vast bodies of research on how pregnancy affects many bodily systems, from research on preeclampsia and gestational diabetes to that on the behavior of autoimmune disease in pregnancy, and more. We have at least diagnostic (if not predictive) biomarkers for some of those entities (as in elevated glucose for diabetes and protein in the urine for preeclampsia), but very little understanding of how the biological and psychosocial changes of pregnancy and the postpartum create changes in the brain—hence a search for biomarkers that came from the idea of treating perinatal mental illness as an obstetrical morbidity.

This book, then, is not a catalog of existing biomarkers, but rather an attempt to gather together in one place what is currently known about biomarkers that are predictive of or reflective of postpartum psychiatric illness including (but not limited to) postpartum depression, postpartum obsessive compulsive disorder (OCD), and postpartum psychosis. The reader will note that there are large gaps in knowledge—more is known about postpartum depression, for example, than postpartum OCD, and there are whole areas that have just begun to be explored, including immunological and imaging biomarkers. The newer fields of proteomics and metabolomics are only beginning to be applied to perinatal mental illness, and there is not enough research on these areas yet to catalog them here. It is our hope that this work will serve as a "road map" for future researchers to fill in these gaps.

This book would not be possible without the research and writing of all those who have contributed, nor without the efforts of other collaborators whose work is not represented here, the hundreds of research staff members who have helped us accomplish this work, and the thousands of research participants who have volunteered their time and their brains and bodies. Progress in science is a "team sport," and we have been honored to work with such outstanding colleagues. It is our hope that we will revisit this book again in the future and be able to fill in many of the gaps that now exist with data and answers that are sorely needed. Enjoy!

Lauren M. Osborne, MD
Jennifer L. Payne, MD
Editors

References

FDA-NIH Biomarker Working Group. (2016). *BEST (Biomarkers, EndpointS, and other Tools) resource*. Silver Spring (MD): Food and Drug Administration (US).

Lozupone, M., La Montagna, M., D'Urso, F., Daniele, A., Greco, A., Seripa, D., ... Panza, F. (2019). The role of biomarkers in psychiatry. *Advances in Experimental Medicine & Biology, 1118*, 135–162. https://doi.org/10.1007/978-3-030-05542-4_7.

Selleck, M. J., Senthil, M., & Wall, N. R. (2017). Making meaningful clinical use of biomarkers. *Biomarker Insights, 12*. https://doi.org/10.1177/1177271917715236, 1177271917715236.

CHAPTER 1

Clinical phenotypes of peripartum depression and time of onset: the PACT Consortium

Jerry Guintivano[1], Karen T. Putnam[1], Patrick F. Sullivan[1,2,3], Samantha Meltzer-Brody[1]
[1]Department of Psychiatry, University of North Carolina at Chapel Hill, NC, United States of America; [2]Department of Genetics, University of North Carolina at Chapel Hill, NC, United States of America; [3]Department of Medical Epidemiology and Biostatistics, Karolinska Institutet, Stockholm, Sweden

Introduction

The perinatal period is well documented as a high-risk time for onset of psychiatric disorders, including postpartum depression (PPD) and is associated with significant morbidity for mother, infant, and family (Meltzer-Brody et al., 2018). This manifests as an observed increased risk for low birth weight and prematurity, impaired mother—infant attachment, and infant malnutrition during the first year of life (Gavin et al., 2005; Parsons, Young, Rochat, Kringelbach, & Stein, 2012). Perhaps most concerning, maternal suicide in the postpartum period has been identified as a leading cause of maternal mortality (Johannsen et al., 2016). Perinatal depression (PND), broadly defined by the World Health Organization as onset of a major depressive episode during pregnancy or across the first 12 months postpartum, has a lifetime prevalence of 10%—15% (Gavin et al., 2005) in industrialized countries and higher risk in lower-income countries (Sawyer, Ayers, & Smith, 2010). The greatest point prevalence for onset of symptoms is the acute postpartum period (Munk-Olsen, Laursen, Pedersen, Mors, & Mortensen, 2006), but there is growing evidence that many women have onset of symptoms during pregnancy (Biaggi, Conroy, Pawlby, & Pariante, 2016). The public health importance of identifying women who suffer with PND was highlighted by new recommendations of the United States Preventive Services Task Force for depression screening during both pregnancy and postpartum (McKinney, Keyser, Clinton, & Pagliano, 2018; O'Connor, Rossom, Henninger, Groom, & Burda, 2016; O'Connor, Senger, Henninger, Coppola, & Gaynes, 2019).

Biomarkers of Postpartum Psychiatric Disorders
ISBN 978-0-12-815508-0
https://doi.org/10.1016/B978-0-12-815508-0.00001-1

This is also consistent with international guidelines from the United Kingdom (National Institute for Health and Clinical Excellence, 2014), Australia (Australian Government Department of Health, 2013), and the World Health Organization recommendations (World Health Organization, 2015).

The Postpartum Depression: Action Towards Causes and Treatment (PACT; www.pactforthecure.com) Consortium was created to increase our understanding of the heterogeneity of the perinatal period and to address the gaps in knowledge about this important illness. This international research consortium subscribes to successful models of team science, whereby collaboration and sharing of data allows for the attainment of the large sample sizes needed to answer some of the challenging questions facing the field. The PACT Consortium has been focused on elucidating the causes of peripartum psychiatric illness so that tailored prevention is possible and personalized effective treatments are developed. This work is vital given the morbidity associated with PND. Women who experience perinatal psychiatric episodes are at higher mortality risk compared with mothers without psychiatric diagnoses, and the first year after diagnosis represents a time of particularly high relative risk for suicide in this vulnerable group (Johannsen et al., 2016).

Rationale and creation of the PACT Consortium

There is a great need to increase our understanding of the phenotypic differences, course, and prognosis for PND. The past decade has seen significant progress in terms of documentation of the prevalence and has considered "hot button" questions that elicit discussion within the field, namely: (1) is PPD a distinct disorder that is fundamentally different from major depression outside of the perinatal period or is it a subtype of major depression? (2) Is childbirth a trigger for the onset of postpartum psychiatric illness? (3) What is the most scientifically logical approach to define postpartum depression? Specifically, should the definition of PPD refer only to onset of symptoms in the immediate postpartum period, or should the definition be broadened to also include onset of symptoms during pregnancy? (4) Does onset of symptoms during pregnancy constitute a different illness than onset of symptoms in the postpartum period? These are complex questions that require very large sample sizes and also provide a significant impetus for forming a scientific consortium that can collaborate using a team-science approach to conduct novel investigations and increase

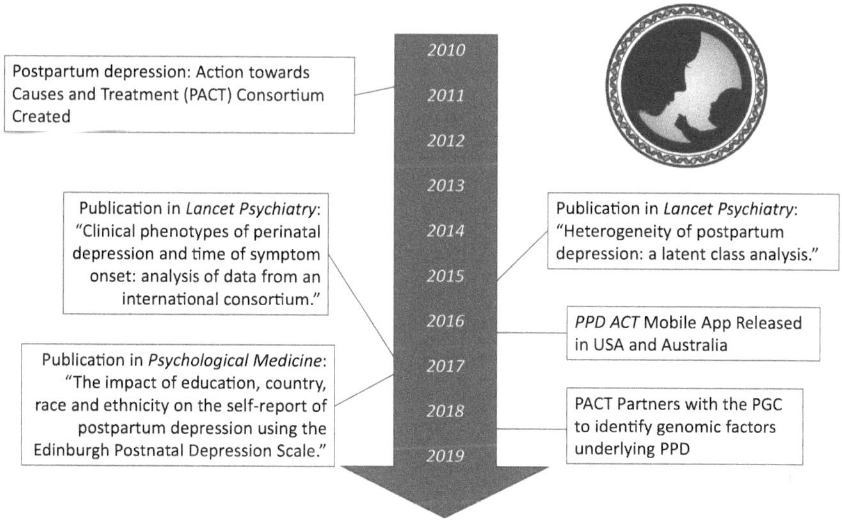

Figure 1.1 Time of PACT milestones.

our understanding about PND. Furthermore, improved understanding of the phenotypic and biological underpinnings of PND may also increase our understanding of the etiology of major depression in general. PPD has the advantage of occurring in reproductive-aged women who give birth and are exposed to a similar biopsychosocial stressor.

The PACT Consortium was formed in 2010 and data analyses to date include investigators from 19 institutions in seven countries. Each investigator contributed previously collected data and agreed to submit this to the PACT Consortium. An overview of PACT accomplishments to date is depicted in Fig. 1.1.

Membership and committees

Membership is an open and democratic process offered to investigators or groups who express interest in contributing anonymized individual-level clinical and/or biological data for analyses. The clinical data are harmonized within PACT with the intent to examine specific PND phenotypes and contain detailed descriptions of the study designs and methods, recruitment, and clinical variables. The biological data include either biospecimens to be used for genomic analyses and/or existing genotyped data with corresponding phenotypic information. Study participants vary by investigator and included perinatal women recruited from a range of clinical settings such as psychiatric clinics, obstetric clinics, primary care, and community advertisements. Individual investigators/studies obtain consent from participants and approval from respective institutional review boards for data sharing. All investigators read and agreed to the rules of conduct described in the PACT memorandum of understanding (MOU) and responded in writing or via e-mail agreeing to follow the terms of the MOU. The PACT MOU was modeled after the successful version created and widely used by the Psychiatric Genomics Consortium (PGC; https://www.med.unc.edu/pgc). The PGC is the largest consortium in the history of psychiatry, with the main purpose of conducting high-quality genomic analyses, and has been very successful. The PACT consortium is following this model and is made up of three committees: the coordinating, phenotyping, and *PPD ACT* committees.

Coordinating committee. The role of the coordinating committee has been to adjudicate issues of relevance to the whole consortium and plan for next steps of PACT. The coordinating committee is meant to primarily facilitate the collaborative efforts of the PACT investigators.

Phenotyping committee. The phenotyping committee has worked to address the phenotypic characterization and determine the common data elements available in the clinical data from PACT investigators. Harmonization of clinical data is a vital step. Prior work has demonstrated that this step must be performed with great rigor or the subsequent interpretation of results becomes unreliable. Given the lack of an international standardized diagnosis of PND, the phenotyping committee had to come up with consensus on how best to operationalize definitions. For example, the *Diagnostic and Statistical Manual of Mental Disorders*, fifth edition (DSM-5), defines PPD as onset of symptoms occurring either during pregnancy or the first month postpartum (American Psychiatric Association & American Psychiatric Association. DSM-5 Task Force, 2013). In contrast, the International Statistical Classification of Diseases, 10th revision (ICD-10), has a more narrow definition of PPD, that is, onset of symptoms within 6 weeks following childbirth. Additionally, the Centers for Disease Control and Prevention and the World Health Organization extend the time of onset of symptoms to the first year postpartum. Thus, the phenotyping committee was tasked with creating a codebook of common and harmonized diagnostic constructs to achieve comparable results across multiple cohorts. This codebook has served a vital role for the work of the PACT Consortium and particularly for the first specific aim.

PPD ACT committee. This committee was created in 2015 to develop and deploy *PPD ACT*, an innovative smartphone study using an app to investigate the genetic signature of PPD (Guintivano et al., 2018). The primary goal for developing the first mobile app for psychiatric genetics research was to have an adequate sample size to power for the first large-scale genome-wide association study (GWAS) for PPD, which is the second specific aim of PACT. There was also great interest in using the power of smartphones to collect data from across the general population. The *PPD ACT* committee carefully designed the app screen by screen, in collaboration with the app developers, to ensure scientific integrity and rigor. In addition, the performance of *PPD ACT* app screening methods was validated clinically (100% sensitivity) and underwent a second test of a subset of cases to demonstrate a high degree of test–retest agreement among cases (85%) (Guintivano et al., 2018). The iOS version of the *PPD ACT* app was successfully launched in March 2016 in the United States and Australia. One year later, we expanded to Canada and have subsequently released an Android and a Spanish version of the app in the United States. Currently, we have identified ~8500 cases, which will contribute to achieving the specific aims of the PACT Consortium.

Approach

A great deal of the data included in the PACT Consortium utilizes clinical diagnosis and participant self-reporting of PPD symptoms. Consequently, we developed and use the lifetime version of the Edinburgh Postnatal Depression Scale (EPDS) (Meltzer-Brody, Boschloo, Jones, Sullivan, & Penninx, 2013) to screen for PPD cases and controls. This version builds on the original EPDS developed by Cox et al. in the United Kingdom, which is one of the most commonly used PND screening instruments across the globe (Boyd, Le, & Somberg, 2005; Cox, Holden, & Sagovsky, 1987; Cox, Murray, & Chapman, 1993; Hewitt & Gilbody, 2009). The EPDS is a 10-item self-report tool that asks about current symptoms, focusing on depression and anxiety including those symptoms that would appear frequently in postpartum women but are not obvious changes in weight or sleep deprivation that come in the immediate postpartum period (Cox et al., 1987). The modified lifetime version of the EPDS asks the participant to focus on the worst time following childbirth in terms of mood and anxiety symptoms. This approach worked very well for our smartphone app that recruited PPD cases, where women who had given birth in their lifetime participate and are screened using the lifetime EPDS. The original EPDS and the lifetime version both follow the standard scoring developed by Cox et al. where total scores fall on scale of 0–30 with higher scores indicating greater symptom severity. EPDS scores ≥13 have been shown to correspond to a clinical diagnosis of PPD (Wisner, Parry, & Piontek, 2002).

Specific research aims of the PACT Consortium

The specific aims of the PACT Consortium were developed to elucidate the heterogeneity of PPD symptoms, including timing of symptom onset and their underlying genetic contributors. Both types of investigations require a team-science approach to allow for the large sample sizes needed to examine subtypes and genetics. The PACT Consortium brought together a large group of investigators to pool resources and collaborate on the specific aims we describe here.

Aim 1. The first aim of the PACT consortium was to identify clinical subtypes by performing a rigorous investigation of the heterogeneity of PPD symptoms. This aim was successfully implemented using existing data from all PACT members, after rigorous data harmonization, as described in the section on the Phenotyping committee. To date, there have been three publications examining this first aim.

The first PACT publication described two latent class analyses (LCAs) of 17,912 individual perinatal women (Postpartum Depression: Action Towards & Treatment Consortium, 2015). LCA is a statistical technique that clusters individuals on the basis of responses in order to create distinct classes. We applied LCA to examine whether PPD could be stratified into empirically defined subtypes based on EPDS responses. The first LCA included all 17,912 individuals, which were clinically defined cases and controls. Subsequently, three unique classes were identified representing (1) those who did not self-report anxiety or depression (mean EPDS score 3.3), (2) those with mild symptoms (mean EPDS score 12.3), and (3) those with more severe symptoms (mean EPDS score 20.3). This approach allowed us to identify both cases (mild and severe) and controls (those with no symptoms). A second LCA was performed on PPD cases ($n = 4245$) to further explore the subtypes of PPD. This LCA was completed using several indicator variables (severity of depressive symptoms, EPDS total score, EPDS anxiety subscale score, complications of pregnancy, obstetric complications, suicidal ideation, and psychiatric history). The result of this PPD case–only analysis demonstrated three distinct classes characterized by severity of symptoms, timing of onset of depressive symptoms, and presence of suicidal thoughts. All classes had high proportions of individuals with psychiatric history.

Upon completion of the first LCA documenting the subtypes of PPD in the PACT consortium cohort, the next step was to further examine these diverse phenotypes, with a particular focus on the type and quality of presentation in women with onset of symptoms during pregnancy compared to postpartum. Based on the literature, we wanted to probe these subtypes using an approach that would account for the underlying pathophysiological mechanisms of PPD, including the hypothesis that hormone fluctuations in vulnerable women may be one of the underlying causes of PPD (Schiller, Meltzer-Brody, & Rubinow, 2015). Consequently, we decided to utilize the symptom constructs of the National Institutes of Mental Health (NIMH) Research Domain Criteria (RDoC) (Insel et al., 2010), which can account for multiple, co-occurring symptom domains. This investigation was a subset of 663 women in the PACT cohort with information about timing of onset of depressive symptom across the perinatal period and complete prospective data including the EPDS (Putnam et al., 2017). Principal components and common factor analyses were used to identify symptom dimensions within the EPDS. The follow-up period was defined as 12 months postpartum split into three categories: 0 to <4 weeks, ≥4 to <8 weeks, and ≥8 weeks. RDoC functional constructs of negative valence and arousal were applied to the EPDS symptom dimensions (depressed mood, anhedonia, and anxiety). Subsequently, using these three dimensions, we identified five subtypes of depression across the perinatal period: severe anxious depression, moderate anxious depression, anxious anhedonia, pure anhedonia, and resolved depression. These subtypes demonstrated clear differences in symptom quality and onset. For example, those classified as having resolved depression typically had onset of symptoms in the second or third trimester of pregnancy and not in the postpartum period. Interestingly, the symptom domains of anxiety and anhedonia were quite robust in those with symptom onset in the immediate postpartum period (<8 weeks). This work demonstrated that unique phenotypes of PPD exist and that it is important to understand and characterize the constellation of symptoms, severity of illness, and timing of onset. This type of investigation sets the stage for future work that will develop tailored treatments based on phenotypic presentations as well as the biological underpinnings of each subtype and is a first step toward personalized medicine within perinatal psychiatry.

Finally, the PACT Consortium was interested in examining the impact of demographics and culture (operationalized as education, ethnicity/race, and continent) on the self-report and experience of postpartum depression using the EPDS (Di Florio et al., 2017). This allowed for an investigation of the role of environment on the heterogeneity of PPD. To examine the role of culture on heterogeneity of PPD, we used a subset of PACT participants ($n = 8209$), based on those with the needed information on ethnicity/race, education, and individual item data on the EPDS. Ordinal regression and measurement invariance were used to explore the association between culture and endorsement of depressive symptoms on the EPDS. Our findings demonstrated that education influenced how PPD is reported and that the structure of EPDS responses was significantly different between Europe and the United States. These findings have important implications for the delivery of personalized clinical care that emphasizes culturally specific and sensitive treatments. Furthermore, this work highlights that education (often a proxy for socioeconomic status) is an important consideration when screening for symptoms and discussing treatment options.

In summary, these three publications to date of the PACT Consortium have allowed for exploration of the complex symptom heterogeneity found within PPD (Di Florio et al., 2017; Postpartum Depression: Action Towards & Treatment Consortium, 2015; Putnam et al., 2017). This approach has been instrumental in efforts to describe a range of symptom severity and classify cases as mild or severe. Furthermore, we have been able to stratify cases using specific symptom domains. Together, this work allows for the identification of specific and clinically relevant subtypes of perinatal psychiatric disorders—a critical step when investigating the underlying biology of the disorder. For example, understanding that anxiety and anhedonia are prominent symptom dimensions with postpartum onset provides crucial insight into the mechanism underlying these symptoms. The postpartum period is characterized by a number of physiological changes (e.g., low levels of reproductive hormones that have rapidly fallen at the time of childbirth), which may trigger severe anxiety and anhedonia in vulnerable women. The postpartum period is markedly different than the high levels of reproductive and stress hormones observed in the second and third trimesters of pregnancy, which was associated with the onset of moderate–severe depression, as well as resolved depression. Thus, the differences in clinical symptoms observed by time of symptom onset could potentially indicate that rising hormones may trigger mood symptoms in some women, but aid in the resolution of symptoms in others. Ultimately, this is a first step and there is much more work needed to understand the biological underpinnings of PND. The next steps will be to replicate our findings and to use these phenotypes to inform a probe of the biological underpinnings of PPD.

Aim 2. The second and current aim of PACT is to elucidate the genetic basis of PPD by conducting a large-scale international GWAS. Interrogating the genetics of PPD is a crucial step into understanding the etiology of the disorder. Genetic variation is often a principal risk factor for psychiatric disorders because exposure begins at the earliest stage of development and can be viewed as the start of a causal pathway that manifests as a complex disorder later in life. The methods used to conduct a GWAS have been described elsewhere and extensively used (Corvin, Craddock, & Sullivan, 2010; Hayes, 2013).

Our understanding of the genetic architecture of many complex phenotypes, particularly psychiatric disorders, continues to evolve. For example, in 2017, the PGC completed the largest metaanalysis for MDD to date (Wray et al., 2018). This effort identified 44 loci that met statistical significance and consistency of effects. In addition, SNPs in genes that encode 42 molecular targets of antidepressant medications (Gaspar & Breen, 2017) were also identified (MAGMA (de Leeuw, Mooij, Heskes, & Posthuma, 2015) gene-enrichment, $P = 8.5 \times 10^{-10}$). This finding was exciting because it connected MDD genomic findings to MDD therapeutics and offers a potential roadmap that may lead to compound discovery for MDD (Breen et al., 2016). This type of large-scale GWAS demonstrated that MDD is not a distinct biological entity but instead can be thought of as a useful clinical construct associated with a range of adverse outcomes and the end result of a complex process of intertwined genetic and environmental effects.

However, in contrast to other disorders, genetic studies for PPD have been sparse, but there is a growing literature. Prior, small genetic epidemiological and linkage studies (See Chapter 2) (Costas et al., 2010; Forty et al., 2006; Mahon et al., 2009; Murphy-Eberenz et al., 2006; Treloar, Martin, Bucholz, Madden, & Heath, 1999) led to the interest of some PACT members to examine the heritability of PPD using the Swedish national twin register, which estimated the heritability of PPD to be 54% and 44% using twin and sibling designs, respectively. In this same population, the heritability of depression occurring outside of the perinatal period was estimated to be 32% (Viktorin et al., 2016). These findings suggest that one-third of the genetic contribution to PND was unique and did not overlap with the genetic component of depression outside of the perinatal period. Another study done by PACT investigators examined the genetic risk for MDD and bipolar disorder in PPD samples (Byrne et al., 2014). This work found empirical genetic evidence for a more important shared genetic etiology between bipolar disorder and PPD than between bipolar disorder and MDD, albeit on small cohorts. Consequently, the increased heritability, as well as increased genetic overlap with other mood disorders, likely makes PPD an exciting candidate for further genetic discovery efforts.

To our knowledge, there has not been a large-scale GWAS performed for PPD using modern genomics techniques. Therefore, the PACT Consortium is focused on conducting the first PPD GWAS and this is currently underway. Work done for other psychiatric GWAS has shown large sample sizes (ideally hundreds of thousands of cases) are required to identify significant risk loci. In order to reach the required sample sizes, samples collected with the *PPD ACT* app (Guintivano et al., 2018) will be combined with existing PACT samples, which are being genotyped to complete this aim. In addition, in order to achieve the very large sample sizes required, we are also partnering with the PGC major depression working group to complete the PPD GWAS. The PGC has conducted and successfully completed many GWAS and secondary analyses which have led to biologically, clinically, and therapeutically meaningful insights across multiple psychiatric disorders. Through the collaboration of the PACT Consortium with the PGC, we will have access to large sample sizes, an infrastructure for analysis, and the expertise to interpret findings in clinically and biologically meaningful ways. Our current goal in this new partnership is to identify specific risk loci with a PPD GWAS, and subsequently use these data to estimate genetic heritability, determine the genetic overlap of PPD with other psychiatric disorders, and generate PPD-specific genetic risk scores. The long-term goal is to stratify subtypes of PPD at the genetic level that can answer important questions in the field, ultimately leading to improved clinical outcomes for the women and their families that suffer with this disorder. The PACT Consortium has a 5-year plan where we expect completion of the first PPD GWAS and an integration of results from identification of PPD subtypes in Aim 1.

Summary and future directions

For the past 8 years, the PACT Consortium has taken great strides into understanding PPD phenomenology using a team-science approach (Fig. 1.1). Our consortium members have recruited 17,912 participants, and

8432 PPD cases have been identified using the *PPD ACT* mobile app. This unique cohort has allowed us to publish three papers detailing heterogeneity among PPD cases. Additionally, we have shown the promise of mobile technology to recruit for genomics studies.

However, there remains much work to be done. Given the prevalence, morbidity, and mortality associated with PPD, an improved approach for detection and treatment is warranted. While we have made significant progress in untangling the heterogeneity of PPD presentations (Aim 1), we need to understand the biological underpinnings including the genetics behind this disorder. Ultimately this could lead to prospective identification of women at risk and change our approach in ways that could lead to prevention. New work by the US Preventive Task Force has identified that psychotherapy could be an effective intervention in the prevention of PND in women identified at risk based on socioeconomic and other demographic factors including trauma history (O'Connor et al., 2019). However, we need to be able to assess biological risk as well as phenotypic risks. Consequently, one of the current goals of the PACT Consortium is to move the bar and implement personalized medicine for peripartum psychiatric disorders similar to that observed in other fields with great success, such as breast cancer (Cardoso et al., 2016; Sparano et al., 2015). Breast cancer researchers have made huge advances by incorporating genomic data with standard clinical practice and this has markedly improved clinical outcomes for many women. Our hope is that as the field of psychiatric genetics continues to increase our knowledge of complex disorders, the work of the PACT Consortium in partnership with the PGC will be well positioned to improve outcomes for women and their families who suffer with PPD and other types of perinatal psychiatric disorders.

References

American Psychiatric Association, & American Psychiatric Association. DSM-5 Task Force. (2013). *Diagnostic and statistical manual of mental disorders: DSM-5* (5th ed.). Washington, D.C.: American Psychiatric Association.

Australian Government Department of Health. (2013). *National perinatal depression initiative*. Commonwealth of Australia. Retrieved from: https://www.health.gov.au/internet/main/publishing.nsf/Content/mental-perinat.

Biaggi, A., Conroy, S., Pawlby, S., & Pariante, C. M. (2016). Identifying the women at risk of antenatal anxiety and depression: A systematic review. *Journal of Affective Disorders, 191*, 62–77. https://doi.org/10.1016/j.jad.2015.11.014.

Boyd, R. C., Le, H. N., & Somberg, R. (2005). Review of screening instruments for postpartum depression. *Archives of Women's Mental Health, 8*(3), 141–153. https://doi.org/10.1007/s00737-005-0096-6.

Breen, G., Li, Q., Roth, B. L., O'Donnell, P., Didriksen, M., Dolmetsch, R., ... Edenberg, H. J. (2016). Translating genome-wide association findings into new therapeutics for psychiatry. *Nature Neuroscience, 19*(11), 1392–1396. https://doi.org/10.1038/nn.4411.

Byrne, E. M., Carrillo-Roa, T., Penninx, B. W., Sallis, H. M., Viktorin, A., Chapman, B., ... Wray, N. R. (2014). Applying polygenic risk scores to postpartum depression. *Archives of Women's Mental Health*. https://doi.org/10.1007/s00737-014-0428-5.

Cardoso, F., van't Veer, L. J., Bogaerts, J., Slaets, L., Viale, G., Delaloge, S., ... MINDACT Investigators. (2016). 70-gene signature as an aid to treatment decisions in early-stage breast cancer. *New England Journal of Medicine, 375*(8), 717–729. https://doi.org/10.1056/NEJMoa1602253.

Corvin, A., Craddock, N., & Sullivan, P. F. (2010). Genome-wide association studies: A primer. *Psychological Medicine, 40*(7), 1063–1077. https://doi.org/10.1017/S0033291709991723.

Costas, J., Gratacos, M., Escaramis, G., Martin-Santos, R., de Diego, Y., Baca-Garcia, E., ... Sanjuan, J. (2010). Association study of 44 candidate genes with depressive and anxiety symptoms in post-partum women. *Journal of Psychiatric Research, 44*(11), 717–724. https://doi.org/10.1016/j.jpsychires.2009.12.012.

Cox, J. L., Holden, J. M., & Sagovsky, R. (1987). Detection of postnatal depression. Development of the 10-item Edinburgh Postnatal Depression Scale. *British Journal of Psychiatry, 150*, 782–786.

Cox, J. L., Murray, D., & Chapman, G. (1993). A controlled study of the onset, duration and prevalence of postnatal depression. *British Journal of Psychiatry, 163*, 27–31.

Di Florio, A., Putnam, K., Altemus, M., Apter, G., Bergink, V., Bilszta, J., ... Meltzer-Brody, S. (2017). The impact of education, country, race and ethnicity on the self-report of postpartum depression using the Edinburgh Postnatal Depression Scale. *Psychological Medicine, 47*(5), 787–799. https://doi.org/10.1017/S0033291716002087.

Forty, L., Jones, L., Macgregor, S., Caesar, S., Cooper, C., Hough, A., ... Jones, I. (2006). Familiality of postpartum depression in unipolar disorder: Results of a family study. *American Journal of Psychiatry, 163*(9), 1549–1553. https://doi.org/10.1176/ajp.2006.163.9.1549.

Gaspar, H. A., & Breen, G. (2017). Pathways analyses of schizophrenia GWAS focusing on known and novel drug targets. *bioRxiv*. https://doi.org/10.1101/091264.

Gavin, N. I., Gaynes, B. N., Lohr, K. N., Meltzer-Brody, S., Gartlehner, G., & Swinson, T. (2005). Perinatal depression: A systematic review of prevalence and incidence. *Obstetrics and Gynecology, 106*(5 Pt 1), 1071–1083.

Guintivano, J., Krohn, H., Lewis, C., Byrne, E. M., Henders, A. K., Ploner, A., ... Meltzer-Brody, S. (2018). PPD ACT: An app-based genetic study of postpartum depression. *Translational Psychiatry, 8*(1), 260. https://doi.org/10.1038/s41398-018-0305-5.

Hayes, B. (2013). Overview of statistical methods for genome-wide association studies (GWAS). *Methods in Molecular Biology, 1019*, 149–169. https://doi.org/10.1007/978-1-62703-447-0_6.

Hewitt, C. E., & Gilbody, S. M. (2009). Is it clinically and cost effective to screen for postnatal depression: A systematic review of controlled clinical trials and economic evidence. *BJOG: An International Journal of Obstetrics and Gynaecology, 116*(8), 1019–1027. https://doi.org/10.1111/j.1471-0528.2009.02148.x.

Insel, T., Cuthbert, B., Garvey, M., Heinssen, R., Pine, D. S., Quinn, K., ... Wang, P. (2010). Research domain criteria (RDoC): Toward a new classification framework for research on mental disorders. *American Journal of Psychiatry, 167*(7), 748–751. https://doi.org/10.1176/appi.ajp.2010.09091379.

Johannsen, B. M., Larsen, J. T., Laursen, T. M., Bergink, V., Meltzer-Brody, S., & Munk-Olsen, T. (2016). All-cause mortality in women with severe postpartum psychiatric disorders. *American Journal of Psychiatry, 173*(6), 635–642. https://doi.org/10.1176/appi.ajp.2015.14121510.

de Leeuw, C. A., Mooij, J. M., Heskes, T., & Posthuma, D. (2015). MAGMA: Generalized gene-set analysis of GWAS data. *PLoS Computational Biology, 11*(4), e1004219. https://doi.org/10.1371/journal.pcbi.1004219.

Mahon, P. B., Payne, J. L., MacKinnon, D. F., Mondimore, F. M., Goes, F. S., Schweizer, B., … Potash, J. B. (2009). Genome-wide linkage and follow-up association study of postpartum mood symptoms. *American Journal of Psychiatry, 166*(11), 1229–1237. https://doi.org/10.1176/appi.ajp.2009.09030417.

McKinney, J., Keyser, L., Clinton, S., & Pagliano, C. (2018). ACOG Committee Opinion No. 736: Optimizing postpartum care. *Obstetrics and Gynecology, 132*(3), 784–785. https://doi.org/10.1097/AOG.0000000000002849.

Meltzer-Brody, S., Boschloo, L., Jones, I., Sullivan, P. F., & Penninx, B. W. (2013). The EPDS-lifetime: Assessment of lifetime prevalence and risk factors for perinatal depression in a large cohort of depressed women. *Archives of Women's Mental Health, 16*(6), 465–473. https://doi.org/10.1007/s00737-013-0372-9.

Meltzer-Brody, S., Howard, L. M., Bergink, V., Vigod, S., Jones, I., Munk-Olsen, T., … Milgrom, J. (2018). Postpartum psychiatric disorders. *Natural Reviews Disease Primers, 4*, 18022. https://doi.org/10.1038/nrdp.2018.22.

Munk-Olsen, T., Laursen, T. M., Pedersen, C. B., Mors, O., & Mortensen, P. B. (2006). New parents and mental disorders: A population-based register study. *The Journal of the American Medical Association, 296*(21), 2582–2589.

Murphy-Eberenz, K., Zandi, P. P., March, D., Crowe, R. R., Scheftner, W. A., Alexander, M., … Levinson, D. F. (2006). Is perinatal depression familial? *Journal of Affective Disorders, 90*(1), 49–55. https://doi.org/10.1016/j.jad.2005.10.006.

National Institute for Health and Clinical Excellence. (2014). *National Collaborating Center for Mental Health. Antenatal and postnatal mental health: The NICE guideline on clinical management and service guidance*. Retrieved from: London, United Kingdom: British Psychological Society.

O'Connor, E., Rossom, R. C., Henninger, M., Groom, H. C., & Burda, B. U. (2016). Primary care screening for and treatment of depression in pregnant and postpartum women: Evidence report and systematic review for the US Preventive Services Task Force. *Journal of the American Medical Association, 315*(4), 388–406. https://doi.org/10.1001/jama.2015.18948.

O'Connor, E., Senger, C. A., Henninger, M. L., Coppola, E., & Gaynes, B. N. (2019). Interventions to prevent perinatal depression: Evidence report and systematic review for the US Preventive Services Task Force. *Journal of the American Medical Association, 321*(6), 588–601. https://doi.org/10.1001/jama.2018.20865.

Parsons, C. E., Young, K. S., Rochat, T. J., Kringelbach, M. L., & Stein, A. (2012). Postnatal depression and its effects on child development: A review of evidence from low- and middle-income countries. *British Medical Bulletin, 101*, 57–79. https://doi.org/10.1093/bmb/ldr047.

Postpartum Depression: Action Towards Causes and Treatment Consortium. (2015). Heterogeneity of postpartum depression: A latent class analysis. *The Lancet Psychiatry, 2*(1), 59–67. https://doi.org/10.1016/S2215-0366(14)00055-8.

Putnam, K. T., Wilcox, M., Robertson-Blackmore, E., Sharkey, K., Bergink, V., Munk-Olsen, T., … Treatment, C. (2017). Clinical phenotypes of perinatal depression and time of symptom onset: Analysis of data from an international consortium. *The Lancet Psychiatry, 4*(6), 477–485. https://doi.org/10.1016/S2215-0366(17)30136-0.

Sawyer, A., Ayers, S., & Smith, H. (2010). Pre- and postnatal psychological wellbeing in Africa: A systematic review. *Journal of Affective Disorders, 123*(1–3), 17–29. https://doi.org/10.1016/j.jad.2009.06.027.

Schiller, C. E., Meltzer-Brody, S., & Rubinow, D. R. (2015). The role of reproductive hormones in postpartum depression. *CNS Spectrums, 20*(1), 48–59. https://doi.org/10.1017/S1092852914000480.
Sparano, J. A., Gray, R. J., Makower, D. F., Pritchard, K. I., Albain, K. S., Hayes, D. F., ... Sledge, G. W. (2015). Prospective validation of a 21-gene expression assay in breast cancer. *New England Journal of Medicine, 373*(21), 2005–2014. https://doi.org/10.1056/NEJMoa1510764.
Treloar, S. A., Martin, N. G., Bucholz, K. K., Madden, P. A., & Heath, A. C. (1999). Genetic influences on post-natal depressive symptoms: Findings from an Australian twin sample. *Psychological Medicine, 29*(3), 645–654.
Viktorin, A., Meltzer-Brody, S., Kuja-Halkola, R., Sullivan, P. F., Landen, M., Lichtenstein, P., & Magnusson, P. K. (2016). Heritability of perinatal depression and genetic overlap with nonperinatal depression. *American Journal of Psychiatry, 173*(2), 158–165. https://doi.org/10.1176/appi.ajp.2015.15010085.
Wisner, K. L., Parry, B. L., & Piontek, C. M. (2002). Clinical practice. Postpartum depression. *New England Journal of Medicine, 347*(3), 194–199. https://doi.org/10.1056/NEJMcp011542.
World Health Organization. (2015). *WHO recommendations on health promotion interventions for maternal and newborn health 2015*. Retrieved from: Geneva, Switzerland: WHO.
Wray, N. R., Ripke, S., Mattheisen, M., Trzaskowski, M., Byrne, E. M., Abdellaoui, A., ... Major Depressive Disorder Working Group of the Psychiatric Genomics Consortium. (2018). Genome-wide association analyses identify 44 risk variants and refine the genetic architecture of major depression. *Nature Genetics, 50*(5), 668–681. https://doi.org/10.1038/s41588-018-0090-3.

CHAPTER 2

Genetic basis for postpartum depression

Jennifer L. Payne[1,2,3]

[1]Department of Psychiatry & Behavioral Sciences, The Johns Hopkins University School of Medicine, Baltimore, MD, United States of America; [2]Department of Gynecology & Obstetrics, The Johns Hopkins University School of Medicine, Baltimore, MD, United States of America; [3]Women's Mood Disorders Center, The Johns Hopkins University School of Medicine, Baltimore, MD, United States of America

Introduction

The search for the genetic basis for various psychiatric disorders has been ongoing for decades, with the hope that identification of genes involved in the development of a psychiatric disorder will shed light on the underlying pathophysiology and potentially lead to improved treatments. Multiple studies indicate that there is likely an underlying genetic vulnerability for postpartum depression (PPD), but it remains unclear if the vulnerability is different from or related to the genetic vulnerability for other mood disorders, including major depressive disorder (MDD) and bipolar disorder. Whether or not PPD is a distinct clinical disorder from other mood disorders also continues to be controversial (Di Florio & Meltzer-Brody, 2015; Silverman, Reichenberg, Lichtenstein, & Sandin, 2018). PPD can be thought of as a major depressive episode (MDE) that begins in the immediate postpartum time period. The postpartum time period appears to confer an elevated risk for the development of a depressive episode, with several large studies demonstrating an increased risk postpartum but not during pregnancy (Di Florio et al., 2013; Eberhard-Gran, Eskild, Tambs, Samuelsen, & Opjordsmoen, 2002; Munk-Olsen, Laursen, Pedersen, Mors, & Mortensen, 2006; Vesga-Lopez et al., 2008). In contrast, a recent study using the Swedish Birth Register calculated the relative risk of PPD in women with a history of depression compared to women without a history of depression using data collected in the year postpartum and a random "phantom date" (Silverman et al., 2018). They concluded that the risk for PPD was no greater in the year following childbirth than following a random date. However, as demonstrated by a previous paper from the same group (Silverman et al., 2017), the risk of PPD rapidly decreases after the

Biomarkers of Postpartum Psychiatric Disorders
ISBN 978-0-12-815508-0
https://doi.org/10.1016/B978-0-12-815508-0.00002-3

first month following delivery. Thus, looking at the risk over the course of a year postpartum may essentially "wash out" the elevated risk in the immediate postpartum period and result in a similar relative risk to time periods outside of the immediate postpartum time period. This paper exemplifies one issue that plagues this literature—timing. Exactly when researchers "measure" the presence or absence of PPD is likely to influence the results of genetic studies, and this fact needs to be kept in mind when interpreting the literature.

Definition of PPD. Turning to another timing issue in the PPD literature, there has been some controversy in exactly how to define PPD. Previously, the most commonly accepted definition has been based on the DSM-IV definition: a MDE that begins within 4 weeks of delivery. The DSM-5 has expanded this definition to include MDEs that begin during pregnancy or within the first 4 weeks after delivery and renamed the condition "MDD with peripartum onset." As discussed in Chapter 1, the onset of symptoms is often, if not more often, during pregnancy with approximately 60% of women who are depressed postpartum reporting onset during pregnancy with symptoms continuing postpartum (PACT, 2015; Wisner et al., 2013). PPD is defined in various ways for the studies described below (as indicated in each description), and the definition used may influence how we should interpret the results.

It remains unclear if there are differences in genetic risk based on timing of onset of symptoms, although the studies that support a genetic basis for PPD have generally defined onset as beginning in the first 4 weeks postpartum (Forty et al., 2006; Murphy-Eberenz et al., 2006; Payne et al., 2008). Episodes that begin in pregnancy and continue postpartum may have a very different biological basis than episodes that clearly began in the immediate postpartum time period. What we call PPD is therefore likely a "complex phenotype" (Di Florio & Meltzer-Brody, 2015) that may consist of several different disease pathways to the illness. Teasing this apart is not an insurmountable task, but will likely take prospective observational studies. Research on the genetic basis of PPD has not only used various definitions but has also used different tools, both self-rated scales and diagnosis based on clinical interviews, to identify PPD. Thus, it is important to consider how "PPD" was defined as we discuss the studies below.

General approach to genetic studies

Before discussing the genetic studies specifically related to PPD, we will give an overview of the types of genetic studies that can be performed in general. The first step to determining the genetic basis of a trait or illness is to establish familiality—or a heritable basis for the trait or illness. Familiality simply demonstrates that the trait or illness runs in families across generations or is more common in close relatives of individuals with the trait or illness. Heritability studies use statistical analyses to estimate how much of a particular trait or illness can be explained by genetics vs. environmental factors. Twin studies—comparing the heritability in identical vs. fraternal twins—are also often used. Traits or illnesses that have a strong genetic basis will occur more commonly in identical twins when compared to fraternal twins. The next level of genetic analyses is to do linkage analysis, in which the specific trait or illness is "linked" to specific areas on chromosomes. The next level is to look as what specific genes are associated with the trait or illness. There are two types of genetic "association" studies: (1) genome-wide association studies, which look for specific genes that are linked or associated with the trait or illness (these often follow a linkage analysis that has identified chromosomes of interest); (2) candidate gene association studies, which identify genes that are thought to have an underlying relationship to the trait or illness of interest and then examine polymorphisms of those genes in populations with and without the trait or illness. A final type of study examines epigenetic variations in genes in association with a trait or illness. Epigenetic studies will be explored in more detail in Chapter 3.

Heritability and familial aggregation of PPD

A number of studies have demonstrated that there is a genetic basis for mood episodes that begin in the immediate postpartum time period, particularly within 4 weeks of delivery. Treloar, Martin, Bucholz, Madden, and Heath (1999) found that genetic factors explained 38% of the variance in PPD as determined by a questionnaire in a sample of 838 twin pairs. Two studies have demonstrated that PPD exhibits familiality in families with MDD, with stronger correlations for depressive episodes that began in the immediate (<4 weeks after delivery) postpartum (Forty et al., 2006; Murphy-Eberenz et al., 2006) as opposed to episodes that began a month or more after delivery. Familiality of PPD has also been in families with bipolar disorder, again when limited to depressive episodes beginning shortly after delivery (Payne et al., 2008). Notably, these studies used retrospective self-reporting and may be subject to recall bias, including regarding the timing of onset of symptoms.

Heritability of PPD versus nonpostpartum MDES

Three studies have attempted to examine whether the genetic basis for PPD is different from the genetic basis for MDEs outside of the postpartum time period. Treloar et al. (1999) found a stronger correlation between PPD and neuroticism than between PPD and nonpostpartum MDEs in the Australian twin registry described above. Byrne et al. (2014) applied polygenic risk scores of bipolar disorder and MDD from the Psychiatric Genomics Consortium datasets in a small sample and found a stronger genetic overlap between bipolar disorder and PPD than between bipolar disorder and nonpostpartum MDEs. Most recently, a Swedish study conducted on 3427 female twins and a population-based cohort of over 500,000 sisters found that the hereditability of PPD is about 40%, with about one-third of this estimate explained by genetic factors not shared with nonpostpartum MDEs (Viktorin et al., 2016). These studies support the hypothesis that the genetic basis for PPD may overlap the genetic basis for other mood disorders but likely also has some unique features.

From a biological perspective it would make sense that at least a portion of the genetic risk for PPD is unique. In an elegant study, Bloch et al. (2000) demonstrated that women with a history of PPD were susceptible to hormonal changes meant to replicate the hormonal shifts that women go through at the time of delivery. Women with a history of PPD developed depressive mood symptoms in response to blinded hormone withdrawal while women without a history of PPD did not. Thus, the women with PPD had a specific mood response to normal hormonal changes, indicating that this vulnerability to hormonal change may be different from the vulnerability to MDD in general.

PPD and bipolar disorder

One other general area deserves mention before moving forward to genetic studies specific to PPD. There are a number of hints in the literature that link PPD to bipolar disorder. For example, a recent study using the Danish birth and psychiatric treatment registers was able to examine the family histories of women who gave birth to their first child from 1970 to 2012 ($N = 362{,}462$) (Bauer et al., 2018). They found that any family history of psychiatric disorders was an important risk factor for the development of postpartum psychiatric disorders, but that the risk was highest with a first-degree family member with bipolar disorder (hazard ratio = 2.86, 95% CI = 1.88–4.35). As noted above, polygenic risk scores for bipolar disorder and MDD from the Psychiatric Genomics Consortium datasets revealed a stronger genetic overlap between bipolar disorder and PPD episodes than between bipolar disorder and nonpostpartum MDEs (Byrne et al., 2014). Finally, a population-based study with 15 years of data collection found a fourfold higher risk of conversion to a bipolar disorder diagnosis in women admitted to the hospital postpartum compared to women admitted during other time periods (Munk-Olsen, Laursen, Meltzer-Brody, Mortensen, & Jones, 2012).

One link between PPD and bipolar disorder is the role of sleep deprivation. Sleep deprivation is thought to play a role in the development of postpartum psychosis and of manic episodes that occur after birth (Sharma & Mazmanian, 2003; Sharma, Smith, & Khan, 2004), and a number of studies have found links between disturbed sleep during pregnancy and postpartum and the development of PPD (Okun, 2016). Thus, a susceptibility to sleep deprivation may underlie the observed association between PPD and bipolar disorder.

Genome-wide linkage and association studies

In a combined linkage analysis and genome-wide association study, Mahon et al. (2009) studied a sample of women with either MDD or bipolar disorder and compared women with PPD episodes (as defined by onset within 4 weeks of delivery) to women who had a history of pregnancy but no PPD. They found suggestive genome-wide linkage for PPD to regions of chromosomes 1 and 9. A follow up single nucleotide polymorphism (SNP) association study of the identified regions found strong association to two intergenic regions but also demonstrated modest association with HMCN1 (hemicentin 1) and METTL13 (methyltransferase like 13) on chromosome 1. These findings were not significant after correction for multiple testing. HMCN1 is highly expressed in the hippocampus, has been shown to be altered in rats by the postpartum drop in estrogen levels (Green & Galea, 2008), and contains four experimentally determined estrogen binding sites (Carroll et al., 2006). METTL13 is putatively involved with methyltransferase activity, which has been shown to play a role in estrogen receptor—induced gene transcription (Green & Galea, 2008).

Mehta et al. (2014) performed a genome-wide association study using peripheral blood gene expression profiles in euthymic pregnant women, comparing the expression profiles during pregnancy between women who went on to develop PPD (defined as onset within the first 7 weeks postpartum) and those who remained euthymic. They identified 116 transcripts that were differentially expressed between the two groups, allowing 88% accuracy in the prediction of PPD. Data mining approaches indicated an overrepresentation of transcripts linked to estrogen signaling, and the authors concluded that women who went on to develop PPD displayed an increased sensitivity to estrogen signaling.

Candidate gene studies

A number of candidate gene studies have been conducted in PPD and are summarized in Table 2.1. Many of the candidate genes studied are thought

Table 2.1 Genetic association studies of postpartum depression.

References	Definition of PPD	Sample size/ country/ ethnicity	Findings
BDNF			
Figueira et al. (2010)	EPDS ≥ 12 at 8 weeks	227/Brazil/NI	Negative 8 weeks
Comasco, Sylven, Papadopoulos, Oreland, et al. (2011)	EPDS ≥ 12 at 6 and 24 weeks	275/Sweden/ NI	Positive in women with a psychiatric history at 6 weeks Negative 24 weeks
COMT Val158Met polymorphism			
Doornbos et al. (2009)	EPDS at 6 and 12 weeks	89/ Netherlands/ Caucasian	Positive 6 weeks Negative 3 months
Comasco, Sylven, Papadopoulos, Sundstrom-Poromaa, et al. (2011)	EPDS ≥ 12 at 6 weeks and 6 months	275/Sweden/ NI	Positive 6 weeks Negative 6 months
Alvim-Soares et al. (2013)	EPDS ≥ 13 at 8 weeks	116/Brazil/ Caucasian	Positive 8 weeks
Corticotrophin-releasing hormone receptor 1			
Engineer et al. (2013)	EPDS ≥ 10 at 2–8 weeks	140/UK/ Caucasian	Positive 2–8 weeks

Table 2.1 Genetic association studies of postpartum depression.—cont'd

References	Definition of PPD	Sample size/ country/ ethnicity	Findings
Schneider et al. (2014)	EPDS at 24 −36 weeks	361/Germany/ Caucasian	Negative 6–9 months
Stergiakouli et al. (2014)	EPDS at 8 weeks	8340/UK/NI	Negative 8 weeks
Tan et al. (2015)	Clinical interview, time-period NI	696/ Singapore/ Chinese	Negative, time-period NI
CYP2D6			
Josefsson et al. (2004)	EPDS ≥ 10 at 6–8 and 24 weeks	145/Sweden/ NI	Negative 6–8 weeks Negative 6 months
Dopamine receptor type 4			
Ivorra et al. (2010)	EPDS at 8 weeks	317/Spain/NI	Negative 8 weeks
ESR-1(α)			
Costas et al. (2010)	EPDS/DIGS at <1, 8, and 32 weeks	1804/Spain/NI 162 women met DSM-IV criteria for an MDE in the first 32 weeks	Positive (but not statistically significant) within 32 weeks
Pinsonneault et al. (2013)	EPDS and/or MADRS, within 12 weeks	156/Canada/ primarily Caucasian	Positive, within 12 weeks
Fatty acid desaturase gene polymorphisms			
Xie and Innis (2009)	EPDS ≥ 10 at 8 and 24 weeks postpartum	69/Canadian/ Caucasian	Negative 8 weeks Positive 6 months
FKBP5			
Schneider et al. (2014)	EPDS at 24–36 weeks	361/Germany/ Caucasian	Negative 6–9 months
Glucocorticoid receptor			
Engineer et al. (2013)	EPDS ≥ 10 at 2–8 weeks	140/UK/ Caucasian	Positive 2–8 weeks

Continued

Table 2.1 Genetic association studies of postpartum depression.—cont'd

References	Definition of PPD	Sample size/ country/ ethnicity	Findings
Schneider et al. (2014)	EPDS at 24–36 weeks	361/Germany/ Caucasian	Negative 6–9 months
Stergiakouli et al. (2014)	EPDS at 8 weeks	8340/UK/NI	Negative 8 weeks
Tan et al. (2015)	Clinical interview, time-period NI	696/ Singapore/ Chinese	Negative, time-period NI
Hemicentin-1 (HMNC1)			
Alvim-Soares et al. (2014)	MINI interview at 8 weeks	110/Brazil/ European	Positive 8 weeks
Hydroxysteroid (11-beta) dehydrogenase 1			
Iliadis et al. (2017)	EPDS at 8 weeks	769/Sweden	Positive 8 weeks Neuroticism played intermediary role
MAOA			
Doornbos et al. (2009)	EPDS at 6 and 12 weeks	89/ Netherlands/ Caucasian	Positive 6 weeks Negative 3 months
Ivorra et al. (2010)	EPDS at 8 weeks	317/Spain/NI	Negative 8 weeks
Comasco, Sylven, Papadopoulos, Sundstrom-Poromaa, et al. (2011)	EPDS ≥ 12 at 6 weeks and 6 months	275/Sweden/ NI	Positive 6 weeks Negative 6 months
Methylenetetra-hydrofolate reductase receptor 1 (MTHFR)			
Lewis et al. (2012)	EPDS at 8–32 weeks and at 8–84 weeks	6809/UK/ Caucasian	Negative 8–32 weeks Positive 8–84 weeks Positive 8 weeks
Oxytocin receptor			
Jonas et al. (2013)	CESD ≥ 27 at 24 weeks	432/Canadian/ Caucasian	Negative 6 months

Table 2.1 Genetic association studies of postpartum depression.—cont'd

References	Definition of PPD	Sample size/ country/ ethnicity	Findings
Mileva-Seitz et al. (2013)	CESD at 24 weeks	187/Canadian/ Caucasian	Positive 6 months
Bell et al. (2015)	EPDS ≥ 12 at 8 weeks	545/UK/ Caucasian	Positive 8 weeks (in mothers not depressed in pregnancy)
Serotonin transporter gene length polymorphism (5-HTTPLRP)			
Sanjuan et al. (2008)	EPDS/DIGS at <1, 8, and 32 weeks	1804/Spain/ Caucasian	Negative <1 week Positive 8 weeks Negative 8 months
Doornbos et al. (2009)	EPDS at 6 and 12 weeks	89/ Netherlands/ Caucasian	Positive 6 weeks Negative 12 weeks
Binder et al. (2010)	SCID and HRSD at <8 weeks, 9–24 weeks	206/EUA/NI	Positive <8 weeks Negative 9–24 weeks
Ivorra et al. (2010)	EPDS at 8 weeks	317/Spain/NI	Negative 8 weeks
Comasco, Sylven, Papadopoulos, Sundstrom-Poromaa, et al (2011)	EPDS ≥ 12 at 6 and 24 weeks	275/Sweden/ NI	Positive in women with a psychiatric history at 6 weeks Negative 6 months
Mitchell et al. (2011)	WHO composite international diagnostic interview based on DSM-IV criteria	1206/US/NI	Negative at 1 year
Gelabert et al. (2012)	SCID and EPDS ≥ 10 at 1–24 weeks	187/Canadian/ Caucasian	Positive (S) 1–24 weeks
Mehta et al. (2014)	EPDS at <1 week and 24–32 weeks	419/German/ Caucasian	Negative <1 week Negative 6–9 months
Khabour et al. (2013)	EPDS > 13 at 4–6 weeks	370/Jordan/NI	Negative 4–6 weeks

Continued

Table 2.1 Genetic association studies of postpartum depression.—cont'd

References	Definition of PPD	Sample size/ country/ ethnicity	Findings
Pinheiro et al. (2013)	MINI plus EPDS ≥ 13 at 6–13 weeks	207/Brazil/NI	Positive 6–13 weeks
Zhang et al. (2014)	Clinical interview and multiple measures including EPDS ≥ 13 in the year postpartum	412/China/ Han Chinese	Positive within the year postpartum
Serotonin transporter VNTR polymorphism			
Sanjuan et al. (2008)	EPDS/DIGS at <1, 8, and 32 weeks	1804/Spain/ Caucasian	Negative <1 week Positive 8 weeks Negative 8 months
Ivorra et al. (2010)	EPDS at 8 weeks	317/Spain/NI	Negative 8 weeks
Mitchell et al. (2011)	WHO composite international diagnostic interview	1206/US/NI	Negative at 1 year
Gelabert et al. (2012)	SCID and EPDS ≥ 10 at 1–24 weeks	187/Canada/ Caucasian	Positive 1–24 weeks
Tryptophan hydroxylase 1 and 2			
Fasching et al. (2012)	EPDS at <1 week and 24–32 weeks	361/Germany/ Caucasian	Negative <1 week Positive 6–8 months
Khabour et al. (2013)	EPDS ≥ 13 at 4–6 weeks	370/Jordan/NI	Negative 4–6 weeks

BDNF, brain-derived neurotrophic factor; *CESD*, Center for Epidemiological Studies Depression Scale; *COMT*, catechol-O-methyltransferase gene; *CYP2D6*, cytochrome P450 2D6 gene; *DIGS*, Diagnostic Interview for Genetic Studies; *EPDS*, Edinburgh Postpartum Depression Scale; *ESR*, estrogen receptor gene; *FKBP5*, FK506 binding protein 5 gene; *HRSD*, Hamilton Rating Scale for Depression; *MAOA*, monoamine oxidase A gene; *MADRS*, Montgomery–Asberg Depression Rating Scale; *MINI*, Mini International Neuropsychiatric Interview; *NI*, not indicated; *SCID*, Structured Clinical Interview for DSM-IV; *VNTR*, variable number of tandem repeats; *WHO*, World Health Organization.

Reprinted with permission from McEvoy, K., Osborne, L. M., Nanavati, J., & Payne, J. L. (2017). Reproductive affective disorders: A review of the genetic evidence for premenstrual dysphoric disorder and postpartum depression. *Current Psychiatry Reports*, *19*(12), 94. doi:10.1007/s11920-017-0852-0, Springer Publishing.

to play a role in MDD in general. In addition, a few candidate genes are thought to play a role in PPD specifically. We review some of the more promising genes below but have included others in Table 2.1 for completeness.

SERT gene. The best studied candidate gene in PPD is the serotonin transporter gene (SERT). The SERT gene has two primary polymorphisms. The first is a 44-bp insertion/deletion located in the promoter region (5-HTTLPR), creating the short and long variants of the gene. The long (L) allele increases transcriptional efficacy and activity compared to the short (S) allele, and the S allele has been associated with mental health problems and depression. The 5-HTTLPR polymorphism is the most extensively studied polymorphism in PPD. The second polymorphism (STin2VNTR) involves a variable number of tandem repeats (VNTR) in the second intron, with longer sets of repeats associated with mental health problems and depression.

There is evidence for a role for SERT polymorphisms in PPD, though studies have had mixed results. The 5-HTTPLRP studies demonstrate a pattern of positive studies when PPD is measured in the immediate postpartum time-period (6–8 weeks) and negative studies when PPD is measured further from delivery (see Table 2.1). As noted, PPD has been shown to be familial, but only when onset of symptoms was in the first 4 weeks after delivery. The 5-HTTPLRP polymorphism may therefore play a role in the susceptibility to PPD, but only in the immediate postpartum time-period. There have been far fewer studies of the STin2VNTR polymorphism in PPD, with mixed results and no clear pattern to date.

COMT gene. The catechol-O-methyltransferase (COMT) gene codes for an enzyme that breaks down catecholamine neurotransmitters including epinephrine, norepinephrine, and dopamine and has previously been studied in depression, anxiety, and stress. The Val158Met polymorphism results in a low activity enzyme in those carrying the Met allele as compared to the Val allele (reviewed in Klein, Schmoeger, Kasper, & Schosser, 2016). There have been conflicting results in studies of this polymorphism in MDD with positive studies with both Val/Val (high activity) and Met/Met (low activity) forms of the gene. In contrast, the COMT low activity (Met/Met) polymorphism has demonstrated a positive association with PPD in three studies—but again only in the immediate postpartum time-period (6–8 weeks) (Alvim-Soares et al., 2013; Comasco, Sylven, Papadopoulos, Sundstrom-Poromaa, et al., 2011; Doornbos et al., 2009), and not when PPD was defined at 3–6 months (Comasco, Sylven, Papadopoulos, Sundstrom-Poromaa, et al., 2011; Doornbos et al., 2009).

MAOA gene. The monoamine oxidase A (MAOA) gene codes for an enzyme that breaks down amine neurotransmitters including serotonin, dopamine, and norepinephrine. MAOA levels have been shown to be elevated in patients with MDD (Meyer et al., 2006), and a high-activity polymorphism of the MAOA gene has been studied in MDD (reviewed in Naoi, Maruyama, & Shamoto-Nagai, 2017) and other psychiatric disorders. There is a functional polymorphism consisting of an untranslated VNTR in the promoter region that is associated with higher activity of the enzyme and, in turn, vulnerability to MDD (Naoi et al., 2017). There have been three studies of the MAOA gene polymorphism in PPD. Two were positive at 6 weeks postpartum but negative at 3 and 6 months postpartum (Comasco, Sylven, Papadopoulos, Sundstrom-Poromaa, et al., 2011; Doornbos et al., 2009), and a third was negative at 8 weeks postpartum (Ivorra et al., 2010).

BDNF gene. Brain-derived neurotrophic factor (BDNF) has been extensively studied in depression. Lower serum levels of the protein have been found in patients with MDD. Notably, serum BDNF levels increase with antidepressant treatment (Lang, Hellweg, & Gallinat, 2004; Shimizu et al., 2003). The Val66Met polymorphism alters the regulated protein secretion, and the Valine allele has been associated with psychiatric disorders. Mice homozygous for the BDNF polymorphism exhibit increased anxiety-like behavior that fluctuated over the estrous cycle (Bath et al., 2012), making BDNF a possible candidate gene for PPD. However, results to date are disappointing. Two studies have been done in women with PPD: one was negative at 8 weeks postpartum (Figueira et al., 2010) and the second was positive at 6 weeks postpartum, but only in women with a previous psychiatric history (Comasco, Sylven, Papadopoulos, Oreland, et al., 2011). The association was negative for PPD measured at 24 weeks, underscoring the importance of timing in gene association studies in PPD.

ESR genes. One hypothesis for the pathophysiology underlying PPD is that the brain generates an abnormal mood response to normal changes in gonadal hormone levels that occur postpartum. As described above, previous work has demonstrated that exposure to and withdrawal from normal levels of gonadal steroids results in depressive symptomatology in women previously diagnosed with PPD (Bloch et al., 2000; Schmidt, Nieman, Danaceau, Adams, & Rubinow, 1998). One possible mechanism for this phenomenon is polymorphic variation in genes encoding gonadal steroid receptors, leading to variability in response to gonadal steroids. The ESR genes are therefore considered candidate genes for PPD. There are two subtypes of estrogen receptors: α (ESR1) and β (ESR2). ESR1 and ESR2 have comparable affinities for estradiol but differ in tissue distribution and other ligand affinities (Westberg & Eriksson, 2008).

Only ESR1 has been studied in PPD. Costas et al. (2010) examined 44 candidate genes for association with PPD in a large sample ($N = 1804$) of Spanish women. Each woman was interviewed at 2–3 days, 8 weeks, and 32 weeks postpartum. The participants initially completed an EPDS, and those who scored >9 were then interviewed with the Diagnostic Interview for Genetics Studies (DIGS), which uses DSM-IV criteria for an MDE. Ultimately 162 women were classified as meeting criteria for PPD within 32 weeks of delivery. In a case–control association study, 4 out of 7 SNPs with a P value less than .01 were from the ESR1 gene. However, no SNP remained significant after correction for multiple tests. A second study (Pinsonneault et al., 2013) enrolled 257 women within 12 weeks of delivery and assessed mood at the time of enrollment with either an EPDS or a Montgomery–Asberg Depression Rating Scale (MADRS), as well as a structured clinical interview (using either the Structured Clinical Interview for DSM-IV [SCID] or the Mini International Neuropsychiatric Interview [MINI]). EPDS scores were available from 156 women and were significantly associated with two ESR1 gene polymorphisms, the TA repeat (L allele) ($P = .007$) and rs2077647 (G allele) ($P = .03$). Only the TA repeat association remained positive after correction for multiple testing ($P = .04$). The same variants were also associated with the clinical diagnosis of PPD, defined as meeting DSM-IV criteria for an MDE within 12 weeks postpartum in women with no previous history of a mood disorder, but this association did not remain significant after correction for multiple tests. These results point to a possible association between ESR1 and PPD and need to be replicated in larger samples. Furthermore, ESR2 has not yet been studied in PPD, and given its role in estrogen signaling it may prove fruitful as well.

OXTR gene. Oxytocin plays a key role in regulating emotion, social interaction, and stress reactivity and is also central to normal birth, lactation, and mother–infant attachment (Carter, 2014; Feldman, 2012). In addition, reductions in oxytocin measured in plasma have been associated with PPD (Skrundz, Bolten, Nast, Hellhammer, & Meinlschmidt, 2011; Stuebe, Grewen, & Meltzer-Brody, 2013). Two studies examined SNP variations in the oxytocin receptor (OXTR) gene and their association with the development of PPD at 24 weeks postpartum and had opposite findings (Jonas et al., 2013; Mileva-Seitz et al., 2013). Interestingly, a more recent study examined variation in the OXTR gene but measured PPD at 8 weeks postpartum and found a positive association (Bell et al., 2015). These findings need to be replicated, and the timing of onset of symptoms should be considered.

HMNC1 gene. As noted above, Mahon et al. (2009) demonstrated a modest association between PPD in women with mood disorders and the hemicentin 1 gene (HMNC1) in a genome-wide linkage and association study. Following up on this finding, Alvim-Soares et al. (2014) genotyped 110 Brazilian women of European descent who completed both an EPDS and an MINI interview at 8 weeks postpartum. They found that heterozygosity at the HMNC1 polymorphism was associated with depressive symptoms. Given that this gene was originally identified by a genome-wide linkage and association study, the confirmation of an association between HMCN1 and PPD is an important finding and needs to be replicated, preferably in larger samples.

Moving forward: approaches for identifyng the genetic basis of PPD

There are several unique initiatives underway with the goal of identifying the genetic basis of PPD. We have described the studies completed to date from the PACT consortium: Postpartum depression Action Towards Causes and Treatment. The PACT consortium is an international group of clinicians and researchers who have compiled independently collected clinical data into a large dataset that has resulted in three publications to date, two of which focused on clinical phenotypes of PPD. A latent class analysis revealed several distinct phenotypes of PPD, including a more severe form that began in pregnancy and continued postpartum and a less severe form that began within 4 weeks of delivery and was associated with pregnancy complications (PACT, 2015). The second publication confirmed the heterogeneity in symptomatology and timing of onset and included both anxious and anhedonic forms of the illness (Putnam et al., 2017). Members of this same group developed the "PPD ACT" mobile application, which when downloaded institutes a basic eligibility screen and an informed consent for PPD phenotyping followed by phenotyping questions (Guintivano et al., 2018). Women who are identified as having a lifetime history of PPD are invited to take part in the DNA collection portion of the study and are sent spit collection kits. This app has allowed the collection of a large sample of DNA from women who met screening criteria for PPD ($N = 2946$ in the first year of collection). The app has also been deployed internationally. Large-scale efforts using such technically savvy techniques will likely expand our ability to identify genetic risk factors for PPD and allow the parsing out of different subtypes of PPD, which may allow for easier identification of involved genes.

Conclusions

It appears that a fairly large proportion (approximately 40%) of the risk for PPD likely has a genetic basis. The genetic risk for PPD may have a component that overlaps with the genetic risk for MDD and/or bipolar disorder and, in addition, a component that is specific to PPD itself. Studies that are attempting to identify genetic risk factors for PPD are complicated by definitional issues, including when to "measure" the presence or absence of PPD, timing of onset of symptoms (during pregnancy, after delivery, how soon after delivery), and differences in measurement tools. Given the high likelihood that the clinical illness we call PPD is heterogenous, prospective studies that carefully measure when a woman becomes symptomatic would potentially be helpful. To date, the candidate gene association studies, as well as the familiality studies, point to a stronger genetic basis when PPD is defined as occurring shortly after birth. Further exploration of genes involved in MDD and bipolar disorder is warranted, as are studies of genes that are responsive to the hormonal changes involved with delivery. Ultimately, identification of the genetic underpinnings of this disorder may shed light on the biological basis for mood disorders more generally.

References

Alvim-Soares, A., Miranda, D., Campos, S. B., Figueira, P., Romano-Silva, M. A., & Correa, H. (2013). Postpartum depression symptoms associated with Val158Met COMT polymorphism. *Archives of Women's Mental Health, 16*(4), 339–340. https://doi.org/10.1007/s00737-013-0349-8.

Alvim-Soares, A. M., Miranda, D. M., Campos, S. B., Figueira, P., Correa, H., & Romano-Silva, M. A. (2014). HMNC1 gene polymorphism associated with postpartum depression. *Revista Brasileira de Psiquiatria, 36*(1), 96–97. https://doi.org/10.1590/1516-4446-2013-3507.

Bath, K. G., Chuang, J., Spencer-Segal, J. L., Amso, D., Altemus, M., McEwen, B. S., & Lee, F. S. (2012). Variant brain-derived neurotrophic factor (Valine66Methionine) polymorphism contributes to developmental and estrous stage-specific expression of anxiety-like behavior in female mice. *Biological Psychiatry, 72*(6), 499–504. https://doi.org/10.1016/j.biopsych.2012.03.032.

Bauer, A. E., Maegbaek, M. L., Liu, X., Wray, N. R., Sullivan, P. F., Miller, W. C., ... Munk-Olsen, T. (2018). Familiality of psychiatric disorders and risk of postpartum psychiatric episodes: A population-based cohort study. *American Journal of Psychiatry, 175*(8), 783–791. https://doi.org/10.1176/appi.ajp.2018.17111184.

Bell, A. F., Carter, C. S., Steer, C. D., Golding, J., Davis, J. M., Steffen, A. D., ... Connelly, J. J. (2015). Interaction between oxytocin receptor DNA methylation and genotype is associated with risk of postpartum depression in women without depression in pregnancy. *Frontiers in Genetics, 6*, 243. https://doi.org/10.3389/fgene.2015.00243.

Binder, E. B., Newport, D. J., Zach, E. B., Smith, A. K., Deveau, T. C., Altshuler, L. L., Cohen, L. S., Stowe, Z. N., & Cubells, J. F. (2010). A serotonin transporter gene polymorphism predicts peripartum depressive symptoms in an at-risk psychiatric cohort. *Journal of Psychiatric Research, 44*(10), 640–646. https://doi.org/10.1016/j.jpsychires.2009.12.001.

Bloch, M., Schmidt, P. J., Danaceau, M., Murphy, J., Nieman, L., & Rubinow, D. R. (2000). Effects of gonadal steroids in women with a history of postpartum depression. *American Journal of Psychiatry, 157*(6), 924–930. https://doi.org/10.1176/appi.ajp.157.6.924.

Byrne, E. M., Carrillo-Roa, T., Penninx, B. W., Sallis, H. M., Viktorin, A., Chapman, B., ... Wray, N. R. (2014). Applying polygenic risk scores to postpartum depression. *Archives of Women's Mental Health, 17*(6), 519–528. https://doi.org/10.1007/s00737-014-0428-5.

Carroll, J. S., Meyer, C. A., Song, J., Li, W., Geistlinger, T. R., Eeckhoute, J., ... Brown, M. (2006). Genome-wide analysis of estrogen receptor binding sites. *Nature Genetics, 38*(11), 1289–1297. https://doi.org/10.1038/ng1901.

Carter, C. S. (2014). Oxytocin pathways and the evolution of human behavior. *Annual Review of Psychology, 65*, 17–39. https://doi.org/10.1146/annurev-psych-010213-115110.

Comasco, E., Sylven, S. M., Papadopoulos, F. C., Oreland, L., Sundstrom-Poromaa, I., & Skalkidou, A. (2011). Postpartum depressive symptoms and the BDNF Val66Met functional polymorphism: Effect of season of delivery. *Archives of Women's Mental Health, 14*(6), 453–463. https://doi.org/10.1007/s00737-011-0239-x.

Comasco, E., Sylven, S. M., Papadopoulos, F. C., Sundstrom-Poromaa, I., Oreland, L., & Skalkidou, A. (2011). Postpartum depression symptoms: A case-control study on monoaminergic functional polymorphisms and environmental stressors. *Psychiatric Genetics, 21*(1), 19–28. https://doi.org/10.1097/YPG.0b013e328341a3c1.

Costas, J., Gratacos, M., Escaramis, G., Martin-Santos, R., de Diego, Y., Baca-Garcia, E., ... Sanjuan, J. (2010). Association study of 44 candidate genes with depressive and anxiety symptoms in post-partum women. *Journal of Psychiatric Research, 44*(11), 717–724. https://doi.org/10.1016/j.jpsychires.2009.12.012.

Di Florio, A., Forty, L., Gordon-Smith, K., Heron, J., Jones, L., Craddock, N., & Jones, I. (2013). Perinatal episodes across the mood disorder spectrum. *JAMA Psychiatry, 70*(2), 168–175. https://doi.org/10.1001/jamapsychiatry.2013.279.

Di Florio, A., & Meltzer-Brody, S. (2015). Is postpartum depression a distinct disorder? *Current Psychiatry Reports, 17*(10), 76. https://doi.org/10.1007/s11920-015-0617-6.

Doornbos, B., Dijck-Brouwer, D. A., Kema, I. P., Tanke, M. A., van Goor, S. A., Muskiet, F. A., & Korf, J. (2009). The development of peripartum depressive symptoms is associated with gene polymorphisms of MAOA, 5-HTT and COMT. *Progress in Neuro-Psychopharmacology & Biological Psychiatry, 33*(7), 1250–1254. https://doi.org/10.1016/j.pnpbp.2009.07.013.

Eberhard-Gran, M., Eskild, A., Tambs, K., Samuelsen, S. O., & Opjordsmoen, S. (2002). Depression in postpartum and non-postpartum women: Prevalence and risk factors. *Acta Psychiatrica Scandinavica, 106*(6), 426–433.

Engineer, N., Darwin, L., Nishigandh, D., Ngianga-Bakwin, K., Smith, S. C., & Grammatopoulos, D. K. (2013). Association of glucocorticoid and type 1 corticotropin-releasing hormone receptors gene variants and risk for depression during pregnancy and post-partum. *Journal of Psychiatric Research, 47*(9), 1166–1173. https://doi.org/10.1016/j.jpsychires.2013.05.003.

Fasching, P. A., Faschingbauer, F., Goecke, T. W., Engel, A., Harberle, L., Seifert, A., & Binder, E. B. (2012). Genetic variants in the tryptophan hydroxylase 2 gene (TPH2) and depression during and after pregnancy. *Journal of Psychiatric Research, 46*(9), 1109–1117. https://doi.org/10.1016.j.jpsychires.2012.05.011.

Feldman, R. (2012). Oxytocin and social affiliation in humans. *Hormones and Behavior, 61*(3), 380–391. https://doi.org/10.1016/j.yhbeh.2012.01.008.

Figueira, P., Malloy-Diniz, L., Campos, S. B., Miranda, D. M., Romano-Silva, M. A., De Marco, L., ... Correa, H. (2010). An association study between the Val66Met polymorphism of the BDNF gene and postpartum depression. *Archives of Women's Mental Health, 13*(3), 285–289. https://doi.org/10.1007/s00737-010-0146-6.

Forty, L., Jones, L., Macgregor, S., Caesar, S., Cooper, C., Hough, A., ... Jones, I. (2006). Familiality of postpartum depression in unipolar disorder: Results of a family study. *American Journal of Psychiatry, 163*(9), 1549–1553. https://doi.org/10.1176/ajp.2006.163.9.1549.

Gelabert, E., Subira, S., Garcia-Esteve, L., Navarro, P., Plaza, A., Cuyas, E., ... Martin-Santos, R. (2012). Perfectionism dimensions in major postpartum depression. *Journal of Affective Disorders, 136*(1-2), 17–25. https://doi.org/10.1016/j.jad.2011.08.030.

Green, A. D., & Galea, L. A. (2008). Adult hippocampal cell proliferation is suppressed with estrogen withdrawal after a hormone-simulated pregnancy. *Hormones and Behavior, 54*(1), 203–211. https://doi.org/10.1016/j.yhbeh.2008.02.023.

Guintivano, J., Krohn, H., Lewis, C., Byrne, E. M., Henders, A. K., Ploner, A., ... Meltzer-Brody, S. (2018). PPD ACT: An app-based genetic study of postpartum depression. *Translational Psychiatry, 8*(1), 260. https://doi.org/10.1038/s41398-018-0305-5.

Iliadis, S. I., Comasco, E., Hellgren, C., Kollia, N., Sundstrom Poromaa, I., & Skalkidou, A. (2017). Associations between a polymorphism in the hydroxysteroid (11-beta) dehydrogenase 1 gene, neuroticism and postpartum depression. *Journal of Affective Disorders, 207*, 141–147. https://doi.org/10.1016/j.jad.2016.09.030.

Ivorra, J. L., Sanjuan, J., Jover, M., Carot, J. M., Frutos, R., & Molto, M. D. (2010). Gene-environment interaction of child temperament. *Journal of Developmental and Behavioral Pediatrics, 31*(7), 545–554. https://doi.org/10.1097/DBP.0b013e3181ee4072.

Jonas, W., Mileva-Seitz, V., Girard, A. W., Bisceglia, R., Kennedy, J. L., Sokolowski, M., ... Steiner, M. (2013). Genetic variation in oxytocin rs2740210 and early adversity associated with postpartum depression and breastfeeding duration. *Genes, Brain and Behavior, 12*(7), 681–694. https://doi.org/10.1111/gbb.12069.

Josefsson, A., Sydsjo, G., Berg, G., Dahl, M. L., Wadelius, M., & Nordin, C. (2004) CYP2D6 genotypes and depressive symptoms during late pregnancy and postpartum. *Nordic Journal of Psychiatry, 58*(1), 61–64. https://doi.org/10.1080/08039480310000815.

Khabour, O., Amarneh, B., Bani Hani, E., & Lataifeh, I. (2013). Associations between variations in TPH1, TPH2, and SCL6A4 genes and postpartum depression: a study in the Jordanian population. *Balkan Journal of Medical Genetics, 16*(1), 41–48. https://doi.org/10.2478/bjmg-2013-0016.

Klein, M., Schmoeger, M., Kasper, S., & Schosser, A. (2016). Meta-analysis of the COMT Val158Met polymorphism in major depressive disorder: The role of gender. *World Journal of Biological Psychiatry, 17*(2), 147–158. https://doi.org/10.3109/15622975.2015.1083615.

Lang, U. E., Hellweg, R., & Gallinat, J. (2004). BDNF serum concentrations in healthy volunteers are associated with depression-related personality traits. *Neuropsychopharmacology, 29*(4), 795–798. https://doi.org/10.1038/sj.npp.1300382.

Lewis, S. J., Araya, R., Leary, S., Smith, G. D., & Ness, A. (2012). Folic acid supplementation during pregnancy may protect against depression 21 months after pregnancy, an effect modified by MTHFR C677T genotype. *European Journal of Clinical Nutrition, 66*(1), 97–103. https://doi.org/10.1038/ejcn.2011.136.

Mahon, P. B., Payne, J. L., MacKinnon, D. F., Mondimore, F. M., Goes, F. S., Schweizer, B., … Potash, J. B. (2009). Genome-wide linkage and follow-up association study of postpartum mood symptoms. *American Journal of Psychiatry, 166*(11), 1229–1237. https://doi.org/10.1176/appi.ajp.2009.09030417.

McEvoy, K., Osborne, L. M., Nanavati, J., & Payne, J. L. (2017). Reproductive affective disorders: A review of the genetic evidence for premenstrual dysphoric disorder and postpartum depression. *Current Psychiatry Reports, 19*(12), 94. https://doi.org/10.1007/s11920-017-0852-0.

Mehta, D., Newport, D. J., Frishman, G., Kraus, L., Rex-Haffner, M., Ritchie, J. C., … Binder, E. B. (2014). Early predictive biomarkers for postpartum depression point to a role for estrogen receptor signaling. *Psychological Medicine, 44*(11), 2309–2322. https://doi.org/10.1017/s0033291713003231.

Meyer, J. H., Ginovart, N., Boovariwala, A., Sagrati, S., Hussey, D., Garcia, A., … Houle, S. (2006). Elevated monoamine oxidase a levels in the brain: An explanation for the monoamine imbalance of major depression. *Archives of General Psychiatry, 63*(11), 1209–1216. https://doi.org/10.1001/archpsyc.63.11.1209.

Mileva-Seitz, V., Steiner, M., Atkinson, L., Meaney, M. J., Levitan, R., Kennedy, J. L., … Fleming, A. S. (2013). Interaction between oxytocin genotypes and early experience predicts quality of mothering and postpartum mood. *PLoS One, 8*(4), e61443. https://doi.org/10.1371/journal.pone.0061443.

Mitchell, C., Notterman, D., Brooks-Gunn, J., Hobcraft, J., Garfinkel, I., Jaeger, K., Kotenko, I., & McLanahan, S. (2011). Role of mother's genes and environment in postpartum depression. *Proceedings of the National Academy of Sciences, 108*(20), 8189–8193. https://doi.org/10.1073/pnas.1014129108.

Munk-Olsen, T., Laursen, T. M., Meltzer-Brody, S., Mortensen, P. B., & Jones, I. (2012). Psychiatric disorders with postpartum onset: Possible early manifestations of bipolar affective disorders. *Archives of General Psychiatry, 69*(4), 428–434. https://doi.org/10.1001/archgenpsychiatry.2011.157.

Munk-Olsen, T., Laursen, T. M., Pedersen, C. B., Mors, O., & Mortensen, P. B. (2006). New parents and mental disorders: A population-based register study. *Journal of the American Medical Association, 296*(21), 2582–2589. https://doi.org/10.1001/jama.296.21.2582.

Murphy-Eberenz, K., Zandi, P. P., March, D., Crowe, R. R., Scheftner, W. A., Alexander, M., … Levinson, D. F. (2006). Is perinatal depression familial? *Journal of Affective Disorders, 90*(1), 49–55. https://doi.org/10.1016/j.jad.2005.10.006.

Naoi, M., Maruyama, W., & Shamoto-Nagai, M. (2017). Type A monoamine oxidase and serotonin are coordinately involved in depressive disorders: From neurotransmitter imbalance to impaired neurogenesis. *Journal of Neural Transmission*. https://doi.org/10.1007/s00702-017-1709-8.

Okun, M. L. (2016). Disturbed sleep and postpartum depression. *Current Psychiatry Reports, 18*(7), 66. https://doi.org/10.1007/s11920-016-0705-2.

PACT. (2015). Heterogeneity of postpartum depression: A latent class analysis. *Lancet Psychiatry, 2*(1), 59–67. https://doi.org/10.1016/s2215-0366(14)00055-8.

Payne, J. L., MacKinnon, D. F., Mondimore, F. M., McInnis, M. G., Schweizer, B., Zamoiski, R. B., … Potash, J. B. (2008). Familial aggregation of postpartum mood symptoms in bipolar disorder pedigrees. *Bipolar Disorders, 10*(1), 38–44. https://doi.org/10.1111/j.1399-5618.2008.00455.x.

Pinheiro, R. T., Coelho, F. M., Silva, R. A., Pinheiro, K. A., Oses, J. P., ... Lucion, A. B. (2013). Association of a serotonin transporter gene polymorphism (5-HTTLPR) and stressful life events with postpartum depressive symptoms: a population-based study. *Journal of Psychosomatic Obstetrics and Gynecology, 34*(1), 29–33. https://doi.org/10.3109/0167482x.2012.759555.

Pinsonneault, J. K., Sullivan, D., Sadee, W., Soares, C. N., Hampson, E., & Steiner, M. (2013). Association study of the estrogen receptor gene ESR1 with postpartum depression – a pilot study. *Archives of Women's Mental Health, 16*(6), 499–509. https://doi.org/10.1007/s00737-013-0373-8.

Putnam, K. T., Wilcox, M., Robertson-Blackmore, E., Sharkey, K., Bergink, V., Munk-Olsen, T., ... Meltzer-Brody, S. (2017). Clinical phenotypes of perinatal depression and time of symptom onset: Analysis of data from an international consortium. *Lancet Psychiatry, 4*(6), 477–485. https://doi.org/10.1016/s2215-0366(17)30136-0.

Sanjuan, J., Martin-Santos, R., Garcia-Esteve, L., Carot, J. M., Guillamat, R., Gutierrez-Zotes, A., ... de Frutos, R. (2008). Mood changes after delivery: role of the serotonin transporter gene. *British Journal of Psychiatry, 193*(5), 383–388. https://doi.org/10.1192/bjp.bp.107.045427.

Schmidt, P. J., Nieman, L. K., Danaceau, M. A., Adams, L. F., & Rubinow, D. R. (1998). Differential behavioral effects of gonadal steroids in women with and in those without premenstrual syndrome. *New England Journal of Medicine, 338*(4), 209–216. https://doi.org/10.1056/nejm199801223380401.

Schneider, M., Engel, A., Fasching, P. A., Haberle, L., Binder, E. B., Voigt, F., & Burghaus, S. (2014). Genetic variants in the genes of the stress hormone signaling pathway and depressive symptoms during and after pregnancy. *BioMed Research International*. https://doi.org/10.1155.2014/469278.

Sharma, V., & Mazmanian, D. (2003). Sleep loss and postpartum psychosis. *Bipolar Disorders, 5*(2), 98–105.

Sharma, V., Smith, A., & Khan, M. (2004). The relationship between duration of labour, time of delivery, and puerperal psychosis. *Journal of Affective Disorders, 83*(2–3), 215–220. https://doi.org/10.1016/j.jad.2004.04.014.

Shimizu, E., Hashimoto, K., Okamura, N., Koike, K., Komatsu, N., Kumakiri, C., ... Iyo, M. (2003). Alterations of serum levels of brain-derived neurotrophic factor (BDNF) in depressed patients with or without antidepressants. *Biological Psychiatry, 54*(1), 70–75.

Silverman, M. E., Reichenberg, A., Lichtenstein, P., & Sandin, S. (2018). Is depression more likely following childbirth? A population-based study. *Archives of Women's Mental Health*. https://doi.org/10.1007/s00737-018-0891-5.

Silverman, M. E., Reichenberg, A., Savitz, D. A., Cnattingius, S., Lichtenstein, P., Hultman, C. M., ... Sandin, S. (2017). The risk factors for postpartum depression: A population-based study. *Depression and Anxiety, 34*(2), 178–187. https://doi.org/10.1002/da.22597.

Skrundz, M., Bolten, M., Nast, I., Hellhammer, D. H., & Meinlschmidt, G. (2011). Plasma oxytocin concentration during pregnancy is associated with development of postpartum depression. *Neuropsychopharmacology, 36*(9), 1886–1893. https://doi.org/10.1038/npp.2011.74.

Stergiakouli, E., Sterne, J. A., & Smith, G. D. (2014). Failure to replicate the association of glucocorticoid and type 1 corticotropin-releasing hormone receptors gene variants with risk of depression during pregnancy and postpartum reported by. *Journal of Psychiatric Research, 56*, 168–170. https://doi.org/10.2016/j.jpsychires.2014.04.016.

Stuebe, A. M., Grewen, K., & Meltzer-Brody, S. (2013). Association between maternal mood and oxytocin response to breastfeeding. *Journal of Women's Health, 22*(4), 352–361. https://doi.org/10.1089/jwh.2012.3768.

Tan, E. C., Chua, T. E., Lee, T. M., Tan, H. S., Ting, J. L., & Chen, H. Y. (2015). Case-control study of glucocorticoid receptor and corticotropin-releasing hormone receptor gene variants and risk of perinatal depression. *BMC Pregnancy and Childbirth, 15*, 283. https://doi.org/10.1186/s12884-015-0720-z.

Treloar, S. A., Martin, N. G., Bucholz, K. K., Madden, P. A., & Heath, A. C. (1999). Genetic influences on post-natal depressive symptoms: Findings from an Australian twin sample. *Psychological Medicine, 29*(3), 645–654.

Vesga-Lopez, O., Schneier, F. R., Wang, S., Heimberg, R. G., Liu, S. M., Hasin, D. S., & Blanco, C. (2008). Gender differences in generalized anxiety disorder: Results from the National Epidemiologic Survey on Alcohol and Related Conditions (NESARC). *Journal of Clinical Psychiatry, 69*(10), 1606–1616.

Viktorin, A., Meltzer-Brody, S., Kuja-Halkola, R., Sullivan, P. F., Landen, M., Lichtenstein, P., & Magnusson, P. K. (2016). Heritability of perinatal depression and genetic overlap with nonperinatal depression. *American Journal of Psychiatry, 173*(2), 158–165. https://doi.org/10.1176/appi.ajp.2015.15010085.

Westberg, L., & Eriksson, E. (2008). Sex steroid-related candidate genes in psychiatric disorders. *Journal of Psychiatry & Neuroscience, 33*(4), 319–330.

Wisner, K. L., Sit, D. K., McShea, M. C., Rizzo, D. M., Zoretich, R. A., Hughes, C. L., … Hanusa, B. H. (2013). Onset timing, thoughts of self-harm, and diagnoses in postpartum women with screen-positive depression findings. *JAMA Psychiatry, 70*(5), 490–498. https://doi.org/10.1001/jamapsychiatry.2013.87.

Xie, L., & Innis, S. M. (2009). Association of fatty acid desaturase gene polymorphisms with blood lipid essential fatty acids and perinatal depression among Canadian women: a pilot study. *Journal of Nutrigenetics and Nutrigenomics, 2*(4-5), 243–250. https://doi.org/10.1159/000255636.

Zhang, A., Wang, L., Huang, F., Li, J., Xiong, L., Xue, H., & Zhang, Y. (2014). Gene-environment interaction in postpartum depression: a Chinese clinical study. *Journal of Affective Disorders, 165*, 208–212. https://doi.org/10.1016/j.jad.2014.04.049.

CHAPTER 3

Epigenetic biomarkers of postpartum depression

Zachary Kaminsky[1], Jennifer L. Payne[2,3,4]
[1]DIFD Mach-Gaensslen Chair of Suicide Prevention Research, The Royal's Institute of Mental Health Research (IMHR), Associate Professor of Cellular and Molecular Medicine, University of Ottawa, Ottawa, Ontario, Canada; [2]Department of Psychiatry & Behavioral Sciences, The Johns Hopkins University School of Medicine, Baltimore, MD, United States of America; [3]Department of Gynecology & Obstetrics, The Johns Hopkins University School of Medicine, Baltimore, MD, United States of America; [4]Women's Mood Disorders Center, The Johns Hopkins University School of Medicine, Baltimore, MD, United States of America

Introduction

Epigenetics is, simply put, Mother Nature's way of turning genes on and off. Epigenetic changes are not modifications in the DNA sequence itself: The Greek prefix "epi" implies features that are "on top of" or "in addition to" the traditional genetic basis for inheritance. Epigenetic modifications determine whether and/or when particular genes are expressed. Every cell in the human body contains the same DNA information and sequence, but some are activated and others turned off, so that each cell type expresses different genes; this is what enables a neuron to function differently from a cardiac cell.

While there is a growing range of modifications that can be termed "epigenetics," the three most well-known and heavily studied include (1) DNA methylation; (2) histone modifications; and (3) noncoding RNA-associated gene silencing. We will not delve into the specific differences here—the important point is that there are different ways to "mark" DNA so that it promotes expression or silencing of that expression for a particular gene. While not necessarily the case for all genes, there is evidence that in some cases, these changes can be heritable and can be passed to future generations. The mechanisms for such a transfer vary and may include the effects of maternal stress and diet during pregnancy on the developing fetus, the establishment of epigenetic marks in early development as a result of parental to infant interactions, and the passage of epigenetic information through the germline. Evidence for the latter is much more difficult to prove causally in humans and remains somewhat controversial. However, both observational studies and mechanistic work in animal models are beginning to shed light in this area. For our purposes it is important to keep

Biomarkers of Postpartum Psychiatric Disorders
ISBN 978-0-12-815508-0
https://doi.org/10.1016/B978-0-12-815508-0.00003-5

in mind that as our understanding of the genetic basis for illness and traits has grown it has become increasingly obvious that epigenetic changes in the expression of genes are likely to underlie many illnesses and that processes that influence epigenetics may, in turn, open up new avenues for treatments.

The most common and well-studied epigenetic process (the one studied in all the research presented in this chapter) is DNA methylation, which involves a chemical modification of the genome by the attachment of methyl groups to cytosines that are primarily located in cytosine-guanine (CpG) sites (Szyf, 2011). These CpG sites are often associated with the transcription start site of the gene, so that increased methylation at these sites generally results in decreased expression of the gene (Jones, 2012).

Perhaps one of the most exciting (and potentially scary) things about epigenetics is the role that environment plays in them. We know that various factors such as diet, stress, pollution, age, and hormones can lead to epigenetic changes in genes, thus leading to changes in gene expression. In psychiatry, we often think of psychiatric illness as a "marriage" between biological vulnerability and environmental stressors—thus the role of epigenetics in psychiatric illness seems like an obvious area for exploration and research. In this chapter, we will discuss epigenetic studies that have been conducted in postpartum depression (PPD). A role for the dramatic hormonal changes that occur during and after delivery in the development of PPD has long been postulated. Bloch et al. elegantly demonstrated that women with a history of PPD have mood changes in response to gonadal hormone fluctuations, while women without a history of PPD do not (Bloch et al., 2000). Hormonally triggered epigenetic changes in gene expression are one possible mechanism underlying PPD. This hypothesis has begun to be explored and results, though mostly preliminary, appear to be promising.

Estrogen-based epigenetic studies in postpartum depression

Our group has published a series of studies identifying two epigenetic biomarkers of PPD. In the first publication (Guintivano, Arad, Gould, Payne, & Kaminsky, 2014), we identified genetic loci that were responsive (i.e., demonstrated significant methylation changes) to high-dose estrogen in the mouse hippocampus. These loci were then cross-referenced to DNA methylation differences in samples of blood from pregnant women with preexisting mood disorders, who were followed prospectively through pregnancy and after delivery. We initially studied women with preexisting mood disorders because they are at higher risk of developing PPD than the general population (Payne et al., 2007; Viguera et al., 2011). The sample used for the epigenetic analyses included pregnant women who were clinically well during pregnancy and either did or did not develop PPD. Using a Linear Discriminant Analysis (LDA) model, we identified two loci at the *HP1BP3* and *TTC9B* genes which, based on methylation differences, were able to correctly identify whether a woman did or did not develop PPD with an 87% accuracy.

We also studied a sample of women who were depressed during pregnancy, a subsample of whom remained depressed postpartum. Given previous work suggesting a genetic vulnerability only in women whose PPD clearly began postpartum, we were interested to see if our biomarkers were or were not predictive of postpartum mood status in women who were depressed during pregnancy. In this sample we found that we could segregate the PPD status with 88% accuracy *but* the prediction was in the opposite direction than in the sample of women who were euthymic during pregnancy and was driven by differences at the *HP1BP3* locus. We found that there was a decrease in the ratio of monocytes to lymphocytes and granulocytes in the antenatally depressed women, and incorporation of these data enabled us to predict the PPD status in the entire cohort with an AUC of 0.82. Ultimately, we determined that incorporation of antenatal mood status (depressed vs. not depressed) even without the blood count data was able to account for the differences in methylation patterns at the *HP1BP3* locus and resulted in a positive predictive value of about 80% (Payne et al., under review).

These results have been replicated in an independent sample of 51 pregnant women with preexisting mood disorders, as well as in a sample of 240 pregnant women participating in the Franconian Maternal Health Evaluation Studies (FRAMES) study who had no previous psychiatric diagnosis, with an accuracy of at least 80% (Osborne et al., 2015). These data provide promising evidence that the epigenetic biomarker patterns at *HP1BP3* and *TTC9B* are replicable in both women with and without preexisting mood disorders.

The exact functions of *TTC9B* and *HP1BP3* are currently unknown, but bioinformatics analysis suggests that both may be involved in mediating synaptic plasticity, and both have ties to estrogen signaling (Guintivano et al., 2014). *TTC9B* has been shown to be responsive to gonadal hormones (Cao, Iyer, & Lin, 2006), and *HP1BP3* associates with estrogen receptor 2 (ESR2) in cell culture (Nassa et al., 2011). A close homologue of *TTC9B*, *TTC9A*, which shares all regulatory motifs, has been demonstrated to regulate estrogen receptor alpha signaling. Additionally, in the context of estrogen supplementation, female *TTC9A* knockout mice exhibit increased anxiety behaviors as well as increased expression of the serotonin precursors Tryptophan hydroxylase1 and 2 and increased expression of the serotonin transporter in the dorsal raphe nucleus (Lim et al., 2016). The authors suggest that *TTC9A* regulates estrogen's effects on anxiety through modulation of the serotonin system. More recently, *HP1BP3* has been implicated in hippocampal synaptic plasticity—mediated cognitive function. *HP1BP3* knockout was shown to instigate neuro-inflammatory processes and decrease the expression of genes involved in synaptic plasticity and neurogenesis in the hippocampus, leading to cognitive decline (Neuner, Ding, & Kaczorowski, 2019). Interestingly, *HP1BP3* knockout mice exhibit impaired maternal care, leading to dramatic reductions in pup survival (Garfinkel et al., 2016). Further work on the functions of these two genes and their role in PPD may ultimately help elucidate the biological underpinnings of PPD.

Since our originally published studies (Guintivano et al., 2014; Osborne et al., 2015), work has focused on moving from our original Linear Discriminant Analysis (LDA) model to a support vector machine (SVM) learning algorithm in order to improve sensitivity and testing replicability and to determine the optimal timing of antenatal blood draw to predict PPD (Payne et al., in press). Both methods are statistical approaches using machine learning to differentiate among two or more groups: here between PPD and no PPD. While LDA focuses on all the data and assumes normality, SVM concentrates on the data that are very close together and difficult to separate into one group or the other, thus improving sensitivity. Our recent work has demonstrated that the SVM algorithm has the ability not only to predict PPD from third trimester blood draws, but also to predict depression later in pregnancy from first trimester blood draws (Payne et al., in press).

In summary, we have identified a biomarker for both antenatal and postpartum mood status that correctly predicts whether or not a woman will be depressed at future points in the perinatal time-period, regardless of whether or not she is depressed at the time of the blood draw, with approximately 80% accuracy. It remains unclear if these epigenetic differences are simply a biomarker for PPD or if they are directly involved in the underlying etiology of PPD. It is conceivable that DNA methylation changes in genes related to hippocampal plasticity are triggered by high levels of estrogen circulating during pregnancy in women susceptible to PPD, and that these biomarkers are therefore one "step" in the pathophysiology underlying PPD. One hypothesis is that the molecular changes exhibited by the biomarkers are indicative of a biological vulnerability, which may interact with various environmental stressors in the postnatal period, ultimately leading to depression.

Since our results also apply to women who were depressed during pregnancy, what does that imply about previous research indicating that there is a genetic basis for episodes of PPD that begin shortly after delivery? PPD is likely to have heterogenous underlying causes, as evidenced by the range of risk factors, which include both biological and environmental factors. The epigenetic changes we have identified may represent a "final common pathway" that indicates that a woman will either remain or become depressed postpartum. The fact that epigenetic methylation patterns were influenced by mood status during pregnancy, with increased methylation at *HP1BP3* in the antepartum depressed sample and decreased methylation at *HP1BP3* in the euthymic sample, indicates that it is the degree of change at this loci that is associated with PPD, not the direction. It will be important to explore this further and identify whether other differences exist between the two populations.

Given these findings, it is important that future research ensures that the populations studied are well defined and, ideally, prospectively collected. It will continue to be important to identify clinical differences in psychiatric populations that may result in "cleaner" biological findings. Had the two populations studied here (the antenatally depressed and antenatally euthymic groups) been studied together, it is likely the biomarkers would not have been identified, given that the methylation changes occurred in the opposite direction. Finally, the utility of these biomarkers as a "test" for PPD is also unclear and remains to be explored. Are we able to intervene in women with the identified biomarkers and prevent onset of PPD or antenatal depression, or do the biomarkers point to an inevitable outcome? Future research will need to focus on prevention strategies for women at high risk. While more research is necessary, the current state of knowledge suggests that prevention is possible, as a comprehensive review of PPD intervention trials demonstrated that numerous factors, including even simply identifying "at-risk" women, can reduce the risk of developing depressive symptomology (Dennis & Dowswell, 2013). If our biomarkers continue to predict peripartum depression with even 80% accuracy, this research will represent a step forward for psychiatry with the development of the first biologically based "test" that would then allow intervention.

Oxytocin receptor epigenetic studies of postpartum depression

There is growing interest in the potential involvement of the oxytocin system in PPD. Oxytocin is a key hormone in the initiation of maternal behavior after parturition (Stuebe, Grewen, Pedersen, Propper, & Meltzer-Brody, 2012), is important for social bonding, and is involved in buffering stress and anxiety. Lower plasma oxytocin concentration levels in midpregnancy have been associated with the later development of PPD (Skrundz, Bolten, Nast, Hellhammer, & Meinlschmidt, 2011), while higher plasma oxytocin levels postpartum predict a higher proportion of positive maternal behaviors (Feldman, Weller, Zagoory-Sharon, & Levine, 2007). Estradiol (E2) increases the rate of transcription of the oxytocin receptor gene (Mamrut et al., 2013), resulting in elevations of oxytocin levels in the uterus (Franczak, Staszkiewicz, Koziorowski, & Kotwica, 2002) and in numerous brain regions, including the hippocampus, the amygdala, and the arcuate nucleus of the hypothalamus (Quinones-Jenab et al., 1997), while genetic variations in the oxytocin receptor may amplify the effects of early life adversity on depression and anxiety (Thompson, Parker, Hallmayer, Waugh, & Gotlib, 2011). This suggests that an interaction between PPD-associated estrogen sensitivity and the oxytocin system may contribute to PPD. Cumulatively, changes in the oxytocin system may occur with past abuse or adversity and be a biological link between history of trauma and risk for antenatal depression and PPD.

A number of studies have investigated oxytocin receptor methylation changes in association with both antenatal and postpartum depression. Higher DNA methylation of the oxytocin receptor gene has been associated with decreased expression of the gene (Kusui et al., 2001). Bell et al. demonstrated an interaction between a specific oxytocin receptor genotype (homozygous at rs53576) and increased DNA methylation at one site in women who were euthymic in pregnancy but developed depression by 8 weeks postpartum (Bell et al., 2015). It is not clear whether the participants had preexisting mood disorders. In contrast, another study found that PPD (defined as onset within 4 weeks postpartum) was linked to *decreased* methylation at two sites in the gene, regardless of genotype (Kimmel et al., 2016). However, this association was only found in women who had a preexisting mood disorder. A third study found that mothers with persistent perinatal depression (i.e., depression during pregnancy and postpartum) had significantly higher methylation of the oxytocin receptor gene at multiple sites (King et al., 2017). Again, it remains unclear whether the participants had preexisting mood disorders. The contrasts among these three studies highlight the importance of the definition of "PPD," which some define as beginning during pregnancy and continuing postpartum and others define as beginning only during the postpartum time-period. In addition, defining whether participants in a PPD study were psychiatrically healthy or had a preexisting mood disorder may be another important variable to consider when examining epigenetic changes in particular genes. In the study by Kimmel et al. (2016) an interaction between DNA methylation variation in the oxytocin receptor and previous adverse life events was observed in association with PPD. There was also a negative correlation between serum estradiol levels and DNA methylation in the oxytocin receptor gene specifically in patients with PPD as well as an interaction between estradiol, oxytocin receptor DNA methylation, and the ratio of allopregnanolone to progesterone (Kimmel et al., 2016). These results circle back to our initial studies, suggesting that the oxytocin receptor may in fact represent a target of increased sensitivity to epigenetic change in response to estrogen. Ultimately, these data highlight the interrelationship between epigenetics and neuroendocrine changes associated with PPD.

Conclusions

Studies on epigenetic modifications in PPD are just emerging, but these early studies demonstrate promise and provide insight regarding the underlying pathophysiological mechanisms, including the roles of estrogen and oxytocin in PPD. Interestingly, these studies point to an interaction between epigenetic modification and signaling by reproductive hormones and neurosteroid levels, bridging multiple mechanisms and bringing together both environmental and biological (genetic) influences. Future work in this area should carefully define the phenotypes of participants including the presence or absence of preexisting mood disorders and the timing of onset of symptoms.

References

Bell, A. F., Carter, C. S., Steer, C. D., Golding, J., Davis, J. M., Steffen, A. D., … Connelly, J. J. (2015). Interaction between oxytocin receptor DNA methylation and genotype is associated with risk of postpartum depression in women without depression in pregnancy. *Frontiers in Genetics, 6*, 243. https://doi.org/10.3389/fgene.2015.00243.

Bloch, M., Schmidt, P. J., Danaceau, M., Murphy, J., Nieman, L., & Rubinow, D. R. (2000). Effects of gonadal steroids in women with a history of postpartum depression. *American Journal of Psychiatry, 157*(6), 924–930. https://doi.org/10.1176/appi.ajp.157.6.924.

Cao, S., Iyer, J. K., & Lin, V. (2006). Identification of tetratricopeptide repeat domain 9, a hormonally regulated protein. *Biochemical and Biophysical Research Communications, 345*(1), 310–317. https://doi.org/10.1016/j.bbrc.2006.04.091.

Dennis, C. L., & Dowswell, T. (2013). Psychosocial and psychological interventions for preventing postpartum depression. *Cochrane Database of Systematic Reviews, 2*, Cd001134. https://doi.org/10.1002/14651858.CD001134.pub3.

Feldman, R., Weller, A., Zagoory-Sharon, O., & Levine, A. (2007). Evidence for a neuroendocrinological foundation of human affiliation: Plasma oxytocin levels across pregnancy and the postpartum period predict mother-infant bonding. *Psychological Science, 18*(11), 965–970. https://doi.org/10.1111/j.1467-9280.2007.02010.x.

Franczak, A., Staszkiewicz, J., Koziorowski, M., & Kotwica, G. (2002). The influence of estradiol and progesterone on the concentrations of uterine oxytocin receptors and plasma PGFM in response to oxytocin in ovariectomized gilts. *Reproduction Nutrition Development, 42*(4), 327–338.

Garfinkel, B. P., Arad, S., Neuner, S. M., Netser, S., Wagner, S., Kaczorowski, C. C., ... Orly, J. (2016). *HP1BP3* expression determines maternal behavior and offspring survival. *Genes, Brain and Behavior, 15*(7), 678–688. https://doi.org/10.1111/gbb.12312.

Guintivano, J., Arad, M., Gould, T. D., Payne, J. L., & Kaminsky, Z. A. (2014). Antenatal prediction of postpartum depression with blood DNA methylation biomarkers. *Molecular Psychiatry, 19*(5), 560–567. https://doi.org/10.1038/mp.2013.62.

Jones, P. A. (2012). Functions of DNA methylation: Islands, start sites, gene bodies and beyond. *Nature Reviews Genetics, 13*(7), 484–492. https://doi.org/10.1038/nrg3230.

Kimmel, M., Clive, M., Gispen, F., Guintivano, J., Brown, T., Cox, O., ... Kaminsky, Z. (2016). Oxytocin receptor DNA methylation in postpartum depression. *Psychoneuroendocrinology, 69*, 150–160. https://doi.org/10.1016/j.psyneuen.2016.04.008.

King, L., Robins, S., Chen, G., Yerko, V., Zhou, Y., Nagy, C., ... Zelkowitz, P. (2017). Perinatal depression and DNA methylation of oxytocin-related genes: A study of mothers and their children. *Hormones and Behavior, 96*, 84–94. https://doi.org/10.1016/j.yhbeh.2017.09.006.

Kusui, C., Kimura, T., Ogita, K., Nakamura, H., Matsumura, Y., Koyama, M., ... Murata, Y. (2001). DNA methylation of the human oxytocin receptor gene promoter regulates tissue-specific gene suppression. *Biochemical and Biophysical Research Communications, 289*(3), 681–686. https://doi.org/10.1006/bbrc.2001.6024.

Lim, L. W., Shrestha, S., Or, Y. Z., Tan, S. Z., Chung, H. H., Sun, Y., ... Lin, V. C. (2016). Tetratricopeptide repeat domain 9A modulates anxiety-like behavior in female mice. *Scientific Reports, 6*, 37568. https://doi.org/10.1038/srep37568.

Mamrut, S., Harony, H., Sood, R., Shahar-Gold, H., Gainer, H., Shi, Y. J., ... Wagner, S. (2013). DNA methylation of specific CpG sites in the promoter region regulates the transcription of the mouse oxytocin receptor. *PLoS One, 8*(2), e56869. https://doi.org/10.1371/journal.pone.0056869. pii:PONE-D-12-33565.

Nassa, G., Tarallo, R., Ambrosino, C., Bamundo, A., Ferraro, L., Paris, O., ... Weisz, A. (2011). A large set of estrogen receptor beta-interacting proteins identified by tandem affinity purification in hormone-responsive human breast cancer cell nuclei. *Proteomics, 11*(1), 159–165. https://doi.org/10.1002/pmic.201000344.

Neuner, S. M., Ding, S., & Kaczorowski, C. C. (2019). Knockdown of heterochromatin protein 1 binding protein 3 recapitulates phenotypic, cellular, and molecular features of aging. *Aging Cell, 18*(1), e12886. https://doi.org/10.1111/acel.12886.

Osborne, L., Clive, M., Kimmel, M., Gispen, F., Guintivano, J., Brown, T., ... Kaminsky, Z. (2015). Replication of epigenetic postpartum depression biomarkers and variation with hormone levels. *Neuropsychopharmacology*. https://doi.org/10.1038/npp.2015.333.

Payne, J. L., Roy, P. S., Murphy-Eberenz, K., Weismann, M. M., Swartz, K. L., McInnis, M. G., ... Potash, J. B. (2007). Reproductive cycle-associated mood symptoms in women with major depression and bipolar disorder. *Journal of Affective Disorders, 99*(1–3), 221–229. https://doi.org/10.1016/j.jad.2006.08.013.

Quinones-Jenab, V., Jenab, S., Ogawa, S., Adan, R. A., Burbach, J. P., & Pfaff, D. W. (1997). Effects of estrogen on oxytocin receptor messenger ribonucleic acid expression in the uterus, pituitary, and forebrain of the female rat. *Neuroendocrinology, 65*(1), 9–17.

Skrundz, M., Bolten, M., Nast, I., Hellhammer, D. H., & Meinlschmidt, G. (2011). Plasma oxytocin concentration during pregnancy is associated with development of postpartum depression. *Neuropsychopharmacology, 36*(9), 1886–1893. https://doi.org/10.1038/npp.2011.74.

Stuebe, A. M., Grewen, K., Pedersen, C. A., Propper, C., & Meltzer-Brody, S. (2012). Failed lactation and perinatal depression: Common problems with shared neuroendocrine mechanisms? *Journal of Women's Health, 21*(3), 264–272. https://doi.org/10.1089/jwh.2011.3083.

Szyf, M. (2011). The early life social environment and DNA methylation: DNA methylation mediating the long-term impact of social environments early in life. *Epigenetics, 6*(8), 971–978. https://doi.org/10.4161/epi.6.8.16793.

Thompson, R. J., Parker, K. J., Hallmayer, J. F., Waugh, C. E., & Gotlib, I. H. (2011). Oxytocin receptor gene polymorphism (rs2254298) interacts with familial risk for psychopathology to predict symptoms of depression and anxiety in adolescent girls. *Psychoneuroendocrinology, 36*(1), 144–147. https://doi.org/10.1016/j.psyneuen.2010.07.003. pii:S0306-4530 (10)00171-X.

Viguera, A. C., Tondo, L., Koukopoulos, A. E., Reginaldi, D., Lepri, B., & Baldessarini, R. J. (2011). Episodes of mood disorders in 2,252 pregnancies and postpartum periods. *American Journal of Psychiatry, 168*(11), 1179–1185. https://doi.org/10.1176/appi.ajp.2011.11010148.

CHAPTER 4

Potential hormonal and neurochemical biomarkers for postpartum depression

Liisa Hantsoo
Department of Psychiatry, The University of Pennsylvania Perelman School of Medicine, Philadelphia, PA, United States of America

Introduction

Postpartum depression (PPD) affects approximately 10% of women worldwide (Gavin et al., 2005; Gaynes et al., 2005; Hübner-Liebermann, Hausner, & Wittmann, 2012). PPD is characterized by symptoms of major depressive disorder (MDD) such as low mood, anxiety, anhedonia, feelings of guilt or worthlessness, and difficulty concentrating (American Psychiatric Association, 2013). According to the *DSM-5* definition (peripartum depression), these symptoms onset during pregnancy or within 1 month of childbirth (American Psychiatric Assn, 2013). The International Classification of Diseases (ICD) expands the timeframe to 6 weeks postchildbirth (WHO 92). Women may also experience depressive symptoms postpartum that do not meet criteria for PPD. Postpartum "blues" refers to low mood, tearfulness, anxiety, insomnia, poor appetite, and irritability generally lasting for a few days in the weeks following delivery, although there are no established diagnostic criteria for this syndrome (Handley, Dunn, Waldron, & Baker, 1980; O'Hara & Wisner, 2014). Given the profound hormonal changes that occur perinatally, much of the research on biological underpinnings of PPD has focused on ovarian steroids such as estradiol and progesterone. While hormonal factors may play an important role in PPD etiology, there are likely other biological contributing factors. In this chapter, we review clinical evidence for potential biomarkers of PPD, including steroid hormones, hypothalamic—pituitary—adrenal (HPA) axis hormones, and monoamine neurotransmitters, and briefly discuss the potential influence of the microbiome.

Biomarkers of Postpartum Psychiatric Disorders
ISBN 978-0-12-815508-0
https://doi.org/10.1016/B978-0-12-815508-0.00004-7

Neuroactive steroid hormones

Progesterone and allopregnanolone

Progesterone is a steroid hormone that plays critical roles during pregnancy, including fetal development and initiation of labor (Vannuccini, Bocchi, Severi, Challis, & Petraglia, 2016). Progesterone is synthesized largely in the ovary and corpus luteum, and by the placenta during pregnancy. Progesterone levels rise exponentially across pregnancy and drop rapidly with parturition (Schock et al., 2016). Its central and peripheral levels are mirrored by those of its metabolite, 3α-hydroxy-5α-pregnan-20-one (allopregnanolone; ALLO) (Luisi et al., 2000). Progesterone and ALLO are synthesized by and act upon receptors in the central nervous system (CNS) and are thus classified as neurosteroids (Rossetti, Cambiasso, Holschbach, & Cabrera, 2016). Specifically, progesterone and ALLO are positive allosteric modulators of the γ-aminobutyric acid type A receptor ($GABA_AR$) (Belelli & Lambert, 2005). GABA is the primary inhibitory neurotransmitter in the CNS, and therefore progesterone metabolites have anxiolytic and sedative-hypnotic effects.

It may be that some women have a hormone-sensitive subtype of PPD (Schiller, Meltzer-Brody,& Rubinow, 2015), which has also been termed "reproductive depression" (Payne, Palmer and Joffe, 2009). Initial studies found that women experiencing depressive symptoms late in the third trimester or in the 3 days following delivery had lower ALLO than euthymic women (Hellgren, Åkerud, Skalkidou, Bäckström, & Sundström-Poromaa, 2014; Nappi et al., 2001). Similarly, in a recent study of women with clinician-confirmed PPD, each additional nanogram per milliliter of peripheral ALLO in the second trimester produced a 63% reduction in risk of developing PPD (Osborne et al., 2017). Lower ALLO levels in nonpregnant women (Andréen et al., 2009) and blunted ALLO response to stress during pregnancy (Crowley et al., 2016) have been associated with negative mood. Women at risk of developing PPD, due to elevated depressive symptoms during pregnancy or a history of PPD or major depression, had higher plasma progesterone and pregnanolone and lower plasma GABA than low-risk participants (Deligiannidis, Kroll-Desrosiers, Mo, et al., 2016). However, plasma ALLO did not differ between women with and without PPD at 9 weeks nor 6 months postpartum, nor did occipital cortex GABA levels at 6 months postpartum differ (Epperson et al., 2006). It may not be the level of neuroactive steroids (NASs) per se, but the dynamics of withdrawal following parturition, that are implicated in PPD (Smith, Shen, Gong, & Zhou, 2007). This also underscores the importance of careful measurement of timing when assessing potential biomarkers. Withdrawing women from estradiol and progesterone in a double-blind trial produced depressive symptoms in women with a history of PPD, but not in controls with no such history (Bloch et al., 2000). The potential role of GABAergic neurosteroids in PPD is supported by results from recent clinical trials of neurosteroid-based compounds. Brexanolone, an injectable ALLO formulation, rapidly reduced depressive symptoms in women with PPD in recent double-blind randomized placebo-controlled trials (Kanes et al., 2017; Meltzer-Brody et al., 2018). Brexanolone had a rapid onset of action, with clinical response occurring within 60 hours of infusion and persisting for up to 30 days (Meltzer-Brody et al., 2018). Given brexanolone's mechanism of action, as a positive allosteric modulator of $GABA_A$ receptors ($GABA_ARs$), these findings also support a potential role for the GABAergic system in PPD. $GABA_AR$ δ and $\gamma 2$ subunits are downregulated in pregnancy to compensate for the elevated levels of progesterone-derived NASs; postpartum, functional $GABA_AR$ numbers increase in alignment with the rapid decline in progesterone-derived steroids that occurs after parturition (Maguire & Mody, 2008). In mice, a deficiency in $GABA_AR$ δ subunits produced depression-like postpartum behaviors, suggesting that poor ability to regulate $GABA_ARs$ perinatally may contribute to risk for PPD (Maguire & Mody, 2008).

Established biomarkers of GABAergic sensitivity are saccadic eye velocity (SEV) and the acoustic startle response (ASR) (Bengtsson et al., 2015; Hantsoo, Golden, Kornfield, Grillon, & Epperson, 2018). SEV measures the velocity of the eye's movement and has been used to measure $GABA_A$-mediated sedation (van Broekhoven, Bäckström, & Verkes, 2006). Although alterations in SEV are found in NAS-related disorders such as premenstrual dysphoric disorder and polycystic ovary syndrome (Hedström et al., 2015; Timby et al., 2016), SEV has not been examined in PPD. ASR, often measured via eyeblink response to a loud noise, is an objective measure of arousal response to an emotional context or stressor (Hantsoo et al., 2018). In pregnancy, there is evidence that ASR is altered, potentially by levels of NASs (Comasco et al., 2016; Hellgren, Bannbers, Åkerud, Risbrough, & Sundström Poromaa, 2012; Kask, Bäckström, Gulinello, & Sundström-Poromaa, 2008). This is consistent with rodent research, in which levels of ALLO modulate ASR (Toufexis, Davis, Hammond, & Davis, 2004). While SEV and ASR have not been studied specifically in PPD, they may be potential biomarkers of susceptibility to PPD, particularly for NAS-mediated subtypes.

Estrogens

The estrogens include estrone, estradiol, and estriol. Estradiol levels increase over pregnancy and drop rapidly postpartum (Schock et al., 2016). Animal studies have suggested that estradiol withdrawal contributes to depressive symptoms, potentially by affecting dopaminergic or serotonergic function (Bethea, Lu, Gundlah, & Streicher, 2002; Byrnes, Byrnes, & Bridges, 2001; Galea, Wide, & Barr, 2001). Similarly in women, those who were administered estradiol and progesterone to mimic pregnancy and then underwent abrupt withdrawal developed depressive symptoms, but only among women with a history of PPD (Bloch et al., 2000). Women who did not have a history of PDD were not sensitive to the NAS withdrawal. Placebo-controlled (Gregoire, Kumar, Everitt, Henderson, & Studd, 1996) and open-label trials (Ahokas, Kaukoranta, Wahlbeck, & Aito, 2001; Sichel, Cohen, Robertson, Ruttenberg, & Rosenbaum, 1995) have improved PPD symptoms with estradiol treatment. This was particularly true in women with low serum levels of estradiol postpartum (Ahokas et al., 2001). Interestingly, women with postpartum depressive symptoms do not show decreased estradiol levels or differences in rate of postpartum estradiol withdrawal compared with controls (Chatzicharalampous et al., 2011; Hohlagschwandtner, Husslein, Klier, & Ulm, 2002; O'Keane et al., 2011; Okun et al., 2011; Stuebe, Grewen, & Meltzer-Brody, 2013). In fact, some studies have found that women with postpartum depressive symptoms have elevated estrogen levels compared with controls (Epperson et al., 2006; Klier et al., 2007). Together, these results suggest that peripheral estrogen levels alone are not a reliable biomarker for PPD risk. It is also possible that differences in timing of sample collection between studies have contributed to mixed findings.

Testosterone

Finally, testosterone has been examined as a potential contributor to PPD. Women with symptoms of PPD had higher serum testosterone levels in the late third trimester (Kallak et al., 2017) and around 24 hours postpartum than women without symptoms (Aswathi et al., 2015). This was also true at term and up to 3 days postpartum (Hohlagschwandtner et al., 2002). However, studies have failed to find associations between peripheral testosterone levels and postpartum depressive symptoms (Chatzicharalampous et al., 2011). Again, these mixed findings indicate that testosterone is not a reliable marker of PPD risk.

The HPA axis

The HPA axis undergoes significant changes during pregnancy. In addition to the pituitary gland, which has increased output, the placenta produces large amounts of corticotropin-releasing hormone (CRH), while hypothalamic CRH is suppressed. Via positive feedback, this results in elevated adrenocorticotropic hormone (ACTH) and cortisol throughout pregnancy (Mastorakos & Ilias, 2003). With delivery of the placenta at parturition, CRH levels drop rapidly. Postpartum, the HPA axis may remain hyporesponsive for several months, particularly among women who are breastfeeding (Lightman et al., 2001; Shanks & Lightman, 2001). In nonpregnant adults, major depression has been associated with abnormalities in HPA axis function (Pariante & Lightman, 2008).

Researchers have proposed that HPA axis dysfunction may similarly play a role in PPD (Glynn, Davis, & Sandman, 2013; Seth, Lewis, & Galbally, 2016). This makes sense, as NASs, which change dramatically across pregnancy and postpartum, are potent regulators of HPA axis function (Brunton & Russell, 2011; Handa & Weiser, 2014). There is a body of evidence supporting the role of the HPA axis in depression during pregnancy and postpartum. In fact, in rodents, chronic corticosterone exposure has been used to model PPD (Brummelte, Pawluski, & Galea, 2006). During pregnancy, elevated placental CRH levels at 25 weeks gestation predicted later development of PPD symptoms (Yim et al., 2009). In healthy women in the second trimester, a blunted ACTH response to acute stress was associated with depressive symptoms (Crowley et al., 2016). At 6 and 12 weeks postpartum, women with depressive symptoms showed decoupling between ACTH and cortisol response to exercise stress (Jolley, Elmore, Barnard, & Carr, 2007). Compared with women lacking depressive symptoms postpartum, those with probable PPD had an elevated cortisol response to acute stress during pregnancy, although no differences in basal cortisol levels (Nierop, Bratsikas, Zimmermann, & Ehlert, 2006). Postpartum blues symptoms were positively associated with ACTH at 5 days postpartum and negatively associated with CRH concentrations during the third trimester (O'Keane et al., 2011). Postpartum depressive symptoms were associated with elevated hair cortisol in the first through third trimesters (Caparros-Gonzalez et al., 2017) and with higher evening salivary cortisol at 6 weeks postpartum (Iliadis et al., 2015) compared to healthy controls. Women with an elevated Edinburgh Postnatal Depression Scale (EPDS) score or past history of PPD or major depression did not differ in their cortisol response to an acute stressor, the Trier Social Stress Test (TSST), compared with healthy women in the third trimester (Deligiannidis, Kroll-Desrosiers, Svenson, et al., 2016). Within a sample of low-income pregnant women, those with HPA abnormalities (including a lower cortisol awakening response [CAR] in the first and second trimester, lower average daily cortisol in the second trimester, a flatter diurnal cortisol pattern in the second and third trimesters and at 3 months postpartum) were more likely to have depressive symptoms at 3 months postpartum (Scheyer & Urizar, 2016). Among breastfeeding women, those with depressive symptoms had an elevated cortisol response to the TSST at 8 weeks postpartum (Cox et al., 2015). Women with PPD had lower morning cortisol at 4 to 6 weeks postpartum (Groer & Morgan, 2007), and blunted CAR at 7 weeks postpartum (Taylor, Glover, Marks, & Kammerer, 2009) compared to women without PPD. Women with PPD had a blunted cortisol response at 4 weeks postpartum and an elevated ACTH response at 8 weeks postpartum to a physical stressor, the cold pressor task (Lara-Cinisomo, Grewen, Girdler, Wood, & Meltzer-Brody, 2017). Women with PPD also had a blunted ACTH response to CRH challenge at 3, 6, and 12 weeks postpartum (Magiakou et al., 1996). Together, these findings suggest impaired negative feedback within the HPA axis. However, a study comparing currently euthymic women with a history of major depression, those with current PPD, and controls found no differences among the groups in HPA response to a pharmacologic nor psychological stressor (Ferguson et al., 2017). Overall, a systemic review of the literature suggested that, even among high-quality studies, the associations between cortisol and PPD are mixed, potentially due to inconsistency in methods (such as timing of sample collection), and further investigation is needed (Seth et al., 2016). Thus, markers of HPA axis activity have promise as potential PPD biomarkers, but additional work is needed.

Thyroid hormones

The thyroid system also undergoes significant changes during pregnancy. Thyroid stimulating hormone (TSH) decreases, and estrogen stimulates the production of thyroid hormone binding globulin, which binds free triiodothyronine (T3) and thyroxine (T4). A compensatory increase in TSH keeps free T3 and T4 levels relatively stable across pregnancy (Moleti, Trimarchi, & Vermiglio, 2014). Findings on potential associations between thyroid hormone and PDD are mixed. Multiple systematic reviews

(Szpunar & Parry, 2018; Yim, Tanner Stapleton, Guardino, Hahn-Holbrook, & Schetter, 2015) did not find clear associations between levels of TSH, T3, or T4 and PPD (Yim et al., 2015). Some studies found an association between elevated TSH or thyroid antibodies postpartum and PPD symptoms (Kuijpens et al., 2001; McCoy et al., 2008; Sylvén et al., 2013), but others did not (Albacar et al., 2010; Le Donne, Settineri, & Benvenga, 2012). Similarly, a few studies found negative associations between T3 or T4 and PPD symptoms (Lambrinoudaki et al., 2010), but others did not (Le Donne et al., 2012; Wissart, Parshad, & Kulkarni, 2005). Some researchers have proposed that it may not be thyroid hormones alone that are implicated in PPD etiology, but thyroid in conjunction with other factors such as estrogens (Yim et al., 2015) or trauma history (Pedersen et al., 2016). At this point, the thyroid hormones are not reliable biomarkers of PPD.

The monoamines

Serotonin

Serotonin (5-HT) is a monoamine neurotransmitter. Alterations in the 5-HT system may contribute to PPD risk. The amino acid tryptophan is a key substrate for 5-HT synthesis, and tryptophan depletion reliably induces symptoms of depression in healthy adults (Ellenbogen, Young, Dean, Palmour, & Benkelfat, 1999; Moreno, McGahuey, Freeman, & Delgado, 2006). Brain tryptophan availability drops following delivery, and this depletion may persist up to 2 months postpartum (Baïlara et al., 2006; Veen et al., 2016), accompanied by decreased platelet serotonin content and increased serotonin transporter (SERT) activity (Baïlara et al., 2006; Doornbos, Fekkes, Tanke, de Jonge, & Korf, 2008). As SERT's role is to remove 5-HT from the synapse, elevated SERT activity may reduce amounts of active 5-HT in postpartum women. In addition, monoamine oxidase A enzyme (MAOA) activity in the brain increases postpartum (Sacher et al., 2010). MAOA is responsible for breakdown of the monoamines, e.g., serotonin, norepinephrine, and dopamine. Increased MAOA binding postpartum reduces monoamines throughout the brain and, in combination with lowered tryptophan availability and estrogen decline, may set the stage for PPD in vulnerable women (Sacher et al., 2010).

The 5-HT system is also influenced by estrogens. Estrogen administration increases 5-HT synthesis (Amin, Canli, & Epperson, 2005; Lokuge, Frey, Foster, Soares, & Steiner, 2010; McQueen, Wilson, & Fink, 1997) and reduces its breakdown by inhibiting MAOA activity (Luine & McEwen, 1977), but also increases SERT levels in the brain (McQueen et al., 1997). Estradiol treatment has been shown to improve SSRI efficacy (Michopoulos, Berga, & Wilson, 2011; Sell et al., 2008). Among women with depressive symptoms postpartum, peripheral 5-HT levels are lower than in women without postpartum depressive symptoms (Xie, Xie, Krewski, & He, 2018; Yildiz et al., 2017). Women with severe PPD exhibited alterations in tryptophan metabolism compared with healthy controls (Veen et al., 2016). Finally, women with PPD had reduced 5HT1A receptor binding compared with healthy postpartum women (Moses-Kolko et al., 2008).

Epinephrine and norepinephrine

Epinephrine and norepinephrine (NE), within the monoamines, are catecholamines that are released from the adrenal glands in response to acute stress. There is relatively little research in the role of epinephrine or norepinephrine in PPD. In one study, epinephrine levels during labor did not correlate with PPD symptoms (Mahomed et al., 1995). Similarly, women with PPD did not differ in epinephrine nor norepinephrine response to pharmacologic nor psychological stressors compared with healthy controls or women with a history of MDD (Ferguson et al., 2017). However, in a case–control study, women scoring in the depression range ($\geq$13) on the EPDS at 2 weeks postpartum had significantly higher plasma levels of NE than controls (Xie et al., 2018).

Dopamine

Dopamine (DA) is also a catecholamine and is involved in reward salience. As reward-related appetitive functions are important in the postpartum period, DA has been one candidate considered in PPD's etiology. In a small sample, women with PPD had a more rapid attenuation of ventral striatal reward response than control mothers (Moses-Kolko et al., 2011). The authors of that study propose that DA-mediated response to reward in areas such as the ventral striatum may contribute to symptoms such as low motivation and goal-directed behavior in PPD (Moses-Kolko et al., 2011). However, in a case–control study, women with probable PPD at 2 weeks postpartum did not show differences in plasma DA levels compared with controls (Xie et al., 2018), nor did DA levels during labor correlate with PPD symptoms (Mahomed et al., 1995). Thus, there are limited data on DA as a potential biomarker for PPD.

Brain-derived Neurotrophic factor

Brain-derived neurotrophic factor (BDNF) is a growth factor within the neurotrophin family that is responsible for the development and support of neurons and synapses, and for neural plasticity. It is believed that stress reduces BDNF levels, increasing risk of major depression in vulnerable individuals (Bocchio-Chiavetto et al., 2010; Brunoni, Lopes, & Fregni, 2008; Molendijk et al., 2014). Serum BDNF levels decrease over the course of pregnancy, then increase postpartum (Christian, Mitchell, Gillespie, & Palettas, 2016), although BDNF levels have shown to be dampened for as long as 2 months postpartum (Lommatzsch et al., 2006). Lower BDNF in the first trimester was associated with depressive symptoms during pregnancy in a cross-sectional study (Fung et al., 2015), and in a longitudinal study low BDNF in the second and third trimesters was associated with depressive symptoms in the third trimester, but not postpartum (Christian et al., 2016). In late pregnancy and at 1 and 8 weeks postpartum, serum BDNF levels were not associated with depression, although this study may have been underpowered due to the small number of participants who developed depression (Lommatzsch et al., 2006). Indeed, in a case–control study, women with probable PPD based on BDI score at 1 to 3 months postpartum had lower serum BDNF levels than controls (Gazal et al., 2012). Thus, BDNF may have promise as a biomarker for PPD, but more research is needed.

The microbiome

Trillions of microorganisms—bacteria, protists, fungi, and viruses—comprise the microbiota of the gut. These microbiota and their genomes are collectively referred to as the gut microbiome. The metabolites produced by the gut microbiota—short chain fatty acids, branched chain fatty acids, branched chain amino acids, and biogenic amines—are the metabolome (Lamichhane, Sen, Dickens, Orešič, & Bertram, 2018). Certain metabolites produced by gut bacteria, such as tryptophan and GABA, play key roles in CNS function and immunomodulation (Barrett, Ross, O'Toole, Fitzgerald, & Stanton, 2012; Dinan & Cryan, 2017a, 2017b; Macfarlane, Cleary, Bahrami, Reynolds, & Macfarlane, 2013). This coordination between the gut, CNS, and neuroendocrine and neuroimmune axes is referred to as the gut–brain axis (Foster & McVey Neufeld, 2013). Adults with major depression show differences in gut microbial community composition compared with controls, and transfer of fecal matter (fecal microbiota transplantation) from depressed humans to microbiota-deficient rats produced depression-like behaviors in the rats (Chen et al., 2018; Kelly et al., 2016; Naseribafrouei et al., 2014).

Little is known to date about the role of the gut–brain axis in PPD. A preliminary report suggested that elevated depressive symptoms in second trimester were associated with reduced gut phylogenetic diversity and reduced species richness, but this did not persist to the third trimester (Togher et al., 2017). In a study of nearly 400 pregnant women who were randomized to a probiotic supplement (*Lactobacillus rhamnosus* HN001) or placebo beginning at 14–16 weeks gestation, those who received *L. rhamnosus* had a lower risk of developing symptoms of depression or anxiety 4 to 8 weeks postpartum than the placebo group (Slykerman et al., 2017). There are several hypothesized pathways via which the gut microbiota could influence risk for depression. One pathway is via neurotransmitters that are produced by the gut microbiota, such as serotonin and GABA, which may influence mood (Barrett et al., 2012; Reigstad et al., 2015; Yunes et al., 2016). Another potential pathway is inflammation. The gut can influence inflammation and immune cell trafficking (Wang et al., 2014), and inflammation has been associated with major depression (Raison, Capuron, & Miller, 2006). During pregnancy, a diverse gut microbiome was associated with less systemic inflammation, although this was not assessed in the context of depression (Röytiö, Mokkala, Vahlberg, & Laitinen, 2017). In sum, the microbiome literature is nascent, and relationships among the gut microbiota, potential mediators such as inflammation or monoamines, and PPD have not been studied in humans to date.

Conclusions

PPD is a complex phenotype, likely influenced by a multitude of risk and resilience factors. While this chapter has touched on potential biological vulnerabilities for PPD, there are also myriad environmental and sociocultural factors that may influence PPD expression. This may include individual factors such as psychiatric history, history of trauma or adverse childhood experiences, current stressors, or material or coping resources (Ahmed, Feng, Bowen, & Muhajarine, 2018; Guintivano et al., 2018; Muzik, Umarji, Sexton, & Davis, 2017). Furthermore, the body of research on PPD biomarkers to date is varied and encompasses a variety of methods for characterizing and assessing PPD. The timing of biomarker measurement during pregnancy or postpartum is a critical issue, with large variation across studies. Ideally, longitudinal studies will assess these factors across time, giving a more precise view of risk than cross-sectional studies. Indeed, the hormonal milieu of pregnancy is complex and dynamic and may interact with vulnerability factors in myriad ways. Thus, development of predictive biomarkers for PPD will be a significant endeavor in the area of women's health research.

References

Ahmed, A., Feng, C., Bowen, A., & Muhajarine, N. (2018). Latent trajectory groups of perinatal depressive and anxiety symptoms from pregnancy to early postpartum and their antenatal risk factors. *Archives of Women's Mental Health*. https://doi.org/10.1007/s00737-018-0845-y.

Ahokas, A., Kaukoranta, J., Wahlbeck, K., & Aito, M. (2001). Estrogen deficiency in severe postpartum depression: Successful treatment with sublingual physiologic 17beta-estradiol: A preliminary study. *The Journal of Clinical Psychiatry, 62*(5), 332–336.

Albacar, G., Sans, T., Martín-Santos, R., García-Esteve, L., Guillamat, R., Sanjuan, J., … Vilella, E. (2010). Thyroid function 48h after delivery as a marker for subsequent postpartum depression. *Psychoneuroendocrinology, 35*(5), 738–742. https://doi.org/10.1016/j.psyneuen.2009.10.015.

American Psychiatric Association. (2013). *Diagnostic and statistical manual of mental disorders* (5th ed.). Arlington VA: American Psychiatric Publishing.

Amin, Z., Canli, T., & Epperson, C. N. (2005). Effect of estrogen-serotonin interactions on mood and cognition. *Behavioral and Cognitive Neuroscience Reviews, 4*(1), 43–58. https://doi.org/10.1177/1534582305277152.

Andréen, L., Nyberg, S., Turkmen, S., van Wingen, G., Fernández, G., & Bäckström, T. (2009). Sex steroid induced negative mood may be explained by the paradoxical effect mediated by GABAA modulators. *Psychoneuroendocrinology, 34*(8), 1121–1132. https://doi.org/10.1016/j.psyneuen.2009.02.003.

Aswathi, A., Rajendiren, S., Nimesh, A., Philip, R. R., Kattimani, S., Jayalakshmi, D., … Dhiman, P. (2015). High serum testosterone levels during postpartum period are associated with postpartum depression. *Asian Journal of Psychiatry, 17*, 85–88. https://doi.org/10.1016/j.ajp.2015.08.008.

Baïlara, K. M., Henry, C., Lestage, J., Launay, J. M., Parrot, F., Swendsen, J., … Demotes-Mainard, J. (2006). Decreased brain tryptophan availability as a partial determinant of post-partum blues. *Psychoneuroendocrinology, 31*(3), 407–413. https://doi.org/10.1016/j.psyneuen.2005.10.001.

Barrett, E., Ross, R. P., O'Toole, P. W., Fitzgerald, G. F., & Stanton, C. (2012). γ-Aminobutyric acid production by culturable bacteria from the human intestine. *Journal of Applied Microbiology, 113*(2), 411–417. https://doi.org/10.1111/j.1365-2672.2012.05344.x.

Belelli, D., & Lambert, J. J. (2005). Neurosteroids: Endogenous regulators of the GABA(A) receptor. *Nature Reviews Neuroscience, 6*(7), 565–575. https://doi.org/10.1038/nrn1703.

Bengtsson, S. K. S., Nyberg, S., Hedström, H., Zingmark, E., Jonsson, B., Bäckström, T., & Bixo, M. (2015). Isoallopregnanolone antagonize allopregnanolone-induced effects on saccadic eye velocity and self-reported sedation in humans. *Psychoneuroendocrinology, 52*, 22–31. https://doi.org/10.1016/j.psyneuen.2014.10.025.

Bethea, C. L., Lu, N. Z., Gundlah, C., & Streicher, J. M. (2002). Diverse actions of ovarian steroids in the serotonin neural system. *Frontiers in Neuroendocrinology, 23*(1), 41–100. https://doi.org/10.1006/frne.2001.0225.

Bloch, M., Schmidt, P. J., Danaceau, M., Murphy, J., Nieman, L., & Rubinow, D. R. (2000). Effects of gonadal steroids in women with a history of postpartum depression. *The American Journal of Psychiatry, 157*(6), 924–930. https://doi.org/10.1176/appi.ajp.157.6.924.

Bocchio-Chiavetto, L., Bagnardi, V., Zanardini, R., Molteni, R., Nielsen, M. G., Placentino, A., … Gennarelli, M. (2010). Serum and plasma BDNF levels in major depression: A replication study and meta-analyses. *The World Journal of Biological Psychiatry: The Official Journal of the World Federation of Societies of Biological Psychiatry, 11*(6), 763–773. https://doi.org/10.3109/15622971003611319.

van Broekhoven, F., Bäckström, T., & Verkes, R. J. (2006). Oral progesterone decreases saccadic eye velocity and increases sedation in women. *Psychoneuroendocrinology, 31*(10), 1190–1199. https://doi.org/10.1016/j.psyneuen.2006.08.007.

Brummelte, S., Pawluski, J. L., & Galea, L. A. M. (2006). High post-partum levels of corticosterone given to dams influence postnatal hippocampal cell proliferation and behavior of offspring: A model of post-partum stress and possible depression. *Hormones and Behavior, 50*(3), 370–382. https://doi.org/10.1016/j.yhbeh.2006.04.008.

Brunoni, A. R., Lopes, M., & Fregni, F. (2008). A systematic review and meta-analysis of clinical studies on major depression and BDNF levels: Implications for the role of neuroplasticity in depression. *The International Journal of Neuropsychopharmacology, 11*(8), 1169–1180. https://doi.org/10.1017/S1461145708009309.

Brunton, P. J., & Russell, J. A. (2011). Allopregnanolone and suppressed hypothalamo-pituitary-adrenal axis stress responses in late pregnancy in the rat. *Stress: The International Journal on the Biology of Stress, 14*(1), 6–12. https://doi.org/10.3109/10253890.2010.482628.

Byrnes, E. M., Byrnes, J. J., & Bridges, R. S. (2001). Increased sensitivity of dopamine systems following reproductive experience in rats. *Pharmacology, Biochemistry and Behavior, 68*(3), 481–489.

Caparros-Gonzalez, R. A., Romero-Gonzalez, B., Strivens-Vilchez, H., Gonzalez-Perez, R., Martinez-Augustin, O., & Peralta-Ramirez, M. I. (2017). Hair cortisol levels, psychological stress and psychopathological symptoms as predictors of postpartum depression. *PLoS One, 12*(8). https://doi.org/10.1371/journal.pone.0182817.

Chatzicharalampous, C., Rizos, D., Pliatsika, P., Leonardou, A., Hasiakos, D., Zervas, I., … Lambrinoudaki, I. (2011). Reproductive hormones and postpartum mood disturbances in Greek women. *Gynecological Endocrinology: The Official Journal of the International Society of Gynecological Endocrinology, 27*(8), 543–550. https://doi.org/10.3109/09513590.2010.501886.

Chen, J., Zheng, P., Liu, Y., Zhong, X., Wang, H., Guo, Y., & Xie, P. (2018). Sex differences in gut microbiota in patients with major depressive disorder. *Neuropsychiatric Disease and Treatment, 14*, 647–655. https://doi.org/10.2147/NDT.S159322.

Christian, L. M., Mitchell, A. M., Gillespie, S. L., & Palettas, M. (2016). Serum brain-derived neurotrophic factor (BDNF) across pregnancy and postpartum: Associations with race, depressive symptoms, and low birth weight. *Psychoneuroendocrinology, 74*, 69–76. https://doi.org/10.1016/j.psyneuen.2016.08.025.

Comasco, E., Gulinello, M., Hellgren, C., Skalkidou, A., Sylven, S., & Sundström-Poromaa, I. (2016). Sleep duration, depression, and oxytocinergic genotype influence prepulse inhibition of the startle reflex in postpartum women. *European Neuropsychopharmacology: The Journal of the European College of Neuropsychopharmacology, 26*(4), 767–776. https://doi.org/10.1016/j.euroneuro.2016.01.002.

Cox, E. Q., Stuebe, A., Pearson, B., Grewen, K., Rubinow, D., & Meltzer-Brody, S. (2015). Oxytocin and HPA stress axis reactivity in postpartum women. *Psychoneuroendocrinology, 55*, 164–172. https://doi.org/10.1016/j.psyneuen.2015.02.009.

Crowley, S. K., O'Buckley, T. K., Schiller, C. E., Stuebe, A., Morrow, A. L., & Girdler, S. S. (2016). Blunted neuroactive steroid and HPA axis responses to stress are associated with reduced sleep quality and negative affect in pregnancy: A pilot study. *Psychopharmacology, 233*(7), 1299–1310. https://doi.org/10.1007/s00213-016-4217-x.

Deligiannidis, K. M., Kroll-Desrosiers, A. R., Mo, S., Nguyen, H. P., Svenson, A., Jaitly, N., ... Shaffer, S. A. (2016). Peripartum neuroactive steroid and γ-aminobutyric acid profiles in women at-risk for postpartum depression. *Psychoneuroendocrinology, 70,* 98–107. https://doi.org/10.1016/j.psyneuen.2016.05.010.

Deligiannidis, K. M., Kroll-Desrosiers, A. R., Svenson, A., Jaitly, N., Barton, B. A., Hall, J. E., & Rothschild, A. J. (2016). Cortisol response to the trier social stress test in pregnant women at risk for postpartum depression. *Archives of Women's Mental Health, 19*(5), 789–797. https://doi.org/10.1007/s00737-016-0615-7.

Dinan, T. G., & Cryan, J. F. (2017a). Gut instincts: Microbiota as a key regulator of brain development, ageing and neurodegeneration. *The Journal of Physiology, 595*(2), 489–503. https://doi.org/10.1113/JP273106.

Dinan, T. G., & Cryan, J. F. (2017b). Microbes, immunity, and behavior: Psychoneuroimmunology meets the microbiome. *Neuropsychopharmacology: Official Publication of the American College of Neuropsychopharmacology, 42*(1), 178–192. https://doi.org/10.1038/npp.2016.103.

Doornbos, B., Fekkes, D., Tanke, M. A. C., de Jonge, P., & Korf, J. (2008). Sequential serotonin and noradrenalin associated processes involved in postpartum blues. *Progress in Neuro-Psychopharmacology and Biological Psychiatry, 32*(5), 1320–1325. https://doi.org/10.1016/j.pnpbp.2008.04.010.

Ellenbogen, M. A., Young, S. N., Dean, P., Palmour, R. M., & Benkelfat, C. (1999). Acute tryptophan depletion in healthy young women with a family history of major affective disorder. *Psychological Medicine, 29*(1), 35–46.

Epperson, C. N., Gueorguieva, R., Czarkowski, K. A., Stiklus, S., Sellers, E., Krystal, J. H., ... Mason, G. F. (2006). Preliminary evidence of reduced occipital GABA concentrations in puerperal women: A 1H-MRS study. *Psychopharmacology, 186*(3), 425–433. https://doi.org/10.1007/s00213-006-0313-7.

Ferguson, E. H., Di Florio, A., Pearson, B., Putnam, K. T., Girdler, S., Rubinow, D. R., & Meltzer-Brody, S. (2017). HPA axis reactivity to pharmacologic and psychological stressors in euthymic women with histories of postpartum versus major depression. *Archives of Women's Mental Health, 20*(3), 411–420. https://doi.org/10.1007/s00737-017-0716-y.

Foster, J. A., & McVey Neufeld, K.-A. (2013). Gut–brain axis: How the microbiome influences anxiety and depression. *Trends in Neurosciences, 36*(5), 305–312. https://doi.org/10.1016/j.tins.2013.01.005.

Fung, J., Gelaye, B., Zhong, Q.-Y., Rondon, M. B., Sanchez, S. E., Barrios, Y. V., ... Williams, M. A. (2015). Association of decreased serum brain-derived neurotrophic factor (BDNF) concentrations in early pregnancy with antepartum depression. *BMC Psychiatry, 15.* https://doi.org/10.1186/s12888-015-0428-7.

Galea, L. A., Wide, J. K., & Barr, A. M. (2001). Estradiol alleviates depressive-like symptoms in a novel animal model of post-partum depression. *Behavioural Brain Research, 122*(1), 1–9.

Gavin, N. I., Gaynes, B. N., Lohr, K. N., Meltzer-Brody, S., Gartlehner, G., & Swinson, T. (2005). Perinatal depression: A systematic review of prevalence and incidence. *Obstetrics and Gynecology, 106*(5 Pt 1), 1071–1083. https://doi.org/10.1097/01.AOG.0000183597.31630.db.

Gaynes, B. N., Gavin, N., Meltzer-Brody, S., Lohr, K. N., Swinson, T., Gartlehner, G., ... Miller, W. C. (2005). Perinatal depression: Prevalence, screening accuracy, and screening outcomes. *Evidence Report/Technology Assessment (Summary)*, (119), 1–8.

Gazal, M., Motta, L. S., Wiener, C. D., Fernandes, J. C., Quevedo, L.Á., Jansen, K., ... Oses, J. P. (2012). Brain-derived neurotrophic factor in post-partum depressive mothers. *Neurochemical Research, 37*(3), 583–587. https://doi.org/10.1007/s11064-011-0647-3.

Glynn, L. M., Davis, E. P., & Sandman, C. A. (2013). New insights into the role of perinatal HPA-axis dysregulation in postpartum depression. *Neuropeptides, 47*(6), 363–370. https://doi.org/10.1016/j.npep.2013.10.007.

Gregoire, A. J., Kumar, R., Everitt, B., Henderson, A. F., & Studd, J. W. (1996). Transdermal oestrogen for treatment of severe postnatal depression. *Lancet (London, England), 347*(9006), 930–933.

Groer, M. W., & Morgan, K. (2007). Immune, health and endocrine characteristics of depressed postpartum mothers. *Psychoneuroendocrinology, 32*(2), 133–139. https://doi.org/10.1016/j.psyneuen.2006.11.007.

Guintivano, J., Sullivan, P. F., Stuebe, A. M., Penders, T., Thorp, J., Rubinow, D. R., & Meltzer-Brody, S. (2018). Adverse life events, psychiatric history, and biological predictors of postpartum depression in an ethnically diverse sample of postpartum women. *Psychological Medicine, 48*(7), 1190–1200. https://doi.org/10.1017/S0033291717002641.

Handa, R. J., & Weiser, M. J. (2014). Gonadal steroid hormones and the hypothalamo-pituitary-adrenal axis. *Frontiers in Neuroendocrinology, 35*(2), 197–220. https://doi.org/10.1016/j.yfrne.2013.11.001.

Handley, S. L., Dunn, T. L., Waldron, G., & Baker, J. M. (1980). Tryptophan, cortisol and puerperal mood. *The British Journal of Psychiatry, 136*(5), 498–508. https://doi.org/10.1192/bjp.136.5.498.

Hantsoo, L., Golden, C. E. M., Kornfield, S., Grillon, C., & Epperson, C. N. (2018). Startling differences: Using the acoustic startle response to study sex differences and neurosteroids in affective disorders. *Current Psychiatry Reports, 20*(6), 40. https://doi.org/10.1007/s11920-018-0906-y.

Hedström, H., Bäckström, T., Bixo, M., Nyberg, S., Wang, M., Gideonsson, I., & Turkmen, S. (2015). Women with polycystic ovary syndrome have elevated serum concentrations of and altered GABAA receptor sensitivity to allopregnanolone. *Clinical Endocrinology*. https://doi.org/10.1111/cen.12809.

Hellgren, C., Åkerud, H., Skalkidou, A., Bäckström, T., & Sundström-Poromaa, I. (2014). Low serum allopregnanolone is associated with symptoms of depression in late pregnancy. *Neuropsychobiology, 69*(3), 147–153. https://doi.org/10.1159/000358838.

Hellgren, C., Bannbers, E., Åkerud, H., Risbrough, V., & Sundström Poromaa, I. (2012). Decreased startle modulation during anticipation in the postpartum period in comparison to late pregnancy. *Archives of Women's Mental Health, 15*(2), 87–94. https://doi.org/10.1007/s00737-012-0261-7.

Hohlagschwandtner, M., Husslein, P., Klier, C., & Ulm, B. (2002). Correlation between serum testosterone levels and peripartal mood states. *Acta Obstetricia et Gynecologica Scandinavica, 80*(4), 326–330. https://doi.org/10.1034/j.1600-0412.2001.080004326.x.

Hübner-Liebermann, B., Hausner, H., & Wittmann, M. (2012). Recognizing and treating peripartum depression. *Deutsches Ärzteblatt International, 109*(24), 419–424. https://doi.org/10.3238/arztebl.2012.0419.

Iliadis, S. I., Comasco, E., Sylvén, S., Hellgren, C., Sundström Poromaa, I., & Skalkidou, A. (2015). Prenatal and postpartum evening salivary cortisol levels in association with peripartum depressive symptoms. *PLoS One, 10*(8), e0135471. https://doi.org/10.1371/journal.pone.0135471.

Jolley, S. N., Elmore, S., Barnard, K. E., & Carr, D. B. (2007). Dysregulation of the hypothalamic-pituitary-adrenal axis in postpartum depression. *Biological Research For Nursing, 8*(3), 210–222. https://doi.org/10.1177/1099800406294598.

Kallak, T. K., Hellgren, C., Skalkidou, A., Sandelin-Francke, L., Ubhayasekhera, K., Bergquist, J., … Sundström Poromaa, I. (2017). Maternal and female fetal testosterone levels are associated with maternal age and gestational weight gain. *European Journal of Endocrinology, 177*(4), 379–388. https://doi.org/10.1530/EJE-17-0207.

Kanes, S., Colquhoun, H., Gunduz-Bruce, H., Raines, S., Arnold, R., Schacterle, A., … Meltzer-Brody, S. (2017). Brexanolone (SAGE-547 injection) in post-partum depression: A randomised controlled trial. *Lancet*. https://doi.org/10.1016/S0140-6736(17)31264-3.

Kask, K., Bäckström, T., Gulinello, M., & Sundström-Poromaa, I. (2008). Lower levels of prepulse inhibition of startle response in pregnant women compared to postpartum women. *Psychoneuroendocrinology, 33*(1), 100–107. https://doi.org/10.1016/j.psyneuen.2007.10.005.

Kelly, J. R., Borre, Y., O' Brien, C., Patterson, E., El Aidy, S., Deane, J., … Dinan, T. G. (2016). Transferring the blues: Depression-associated gut microbiota induces neurobehavioural changes in the rat. *Journal of Psychiatric Research, 82*, 109–118. https://doi.org/10.1016/j.jpsychires.2016.07.019.

Klier, C. M., Muzik, M., Dervic, K., Mossaheb, N., Benesch, T., Ulm, B., & Zeller, M. (2007). The role of estrogen and progesterone in depression after birth. *Journal of Psychiatric Research, 41*(3–4), 273–279. https://doi.org/10.1016/j.jpsychires.2006.09.002.

Kuijpens, J. L., Vader, H. L., Drexhage, H. A., Wiersinga, W. M., van Son, M. J., & Pop, V. J. (2001). Thyroid peroxidase antibodies during gestation are a marker for subsequent depression postpartum. *European Journal of Endocrinology, 145*(5), 579–584.

Lambrinoudaki, I., Rizos, D., Armeni, E., Pliatsika, P., Leonardou, A., Sygelou, A., … Papadias, C. (2010). Thyroid function and postpartum mood disturbances in Greek women. *Journal of Affective Disorders, 121*(3), 278–282. https://doi.org/10.1016/j.jad.2009.07.001.

Lamichhane, S., Sen, P., Dickens, A. M., Orešič, M., & Bertram, H. C. (2018). Gut metabolome meets microbiome: A methodological perspective to understand the relationship between host and microbe. *Methods*. https://doi.org/10.1016/j.ymeth.2018.04.029.

Lara-Cinisomo, S., Grewen, K. M., Girdler, S. S., Wood, J., & Meltzer-Brody, S. (2017). Perinatal depression, adverse life events, and hypothalamic-adrenal-pituitary Axis response to cold pressor stress in Latinas: An exploratory study. *Women's Health Issues: Official Publication of the Jacobs Institute of Women's Health, 27*(6), 673–682. https://doi.org/10.1016/j.whi.2017.06.004

Le Donne, M., Settineri, S., & Benvenga, S. (2012). Early pospartum alexithymia and risk for depression: Relationship with serum thyrotropin, free thyroid hormones and thyroid autoantibodies. *Psychoneuroendocrinology, 37*(4), 519–533. https://doi.org/10.1016/j.psyneuen.2011.08.001.

Lightman, S. L., Windle, R. J., Wood, S. A., Kershaw, Y. M., Shanks, N., & Ingram, C. D. (2001). Peripartum plasticity within the hypothalamo-pituitary-adrenal axis. *Progress in Brain Research, 133*, 111–129.

Lokuge, S., Frey, B. N., Foster, J. A., Soares, C. N., & Steiner, M. (2010). The rapid effects of estrogen: A mini-review. *Behavioural Pharmacology, 21*(5–6), 465–472. https://doi.org/10.1097/FBP.0b013e32833da5c3.

Lommatzsch, M., Hornych, K., Zingler, C., Schuff-Werner, P., Höppner, J., & Virchow, J. C. (2006). Maternal serum concentrations of BDNF and depression in the perinatal period. *Psychoneuroendocrinology, 31*(3), 388–394. https://doi.org/10.1016/j.psyneuen.2005.09.003.

Luine, V. N., & McEwen, B. S. (1977). Effect of oestradiol on turnover of type A monoamine oxidase in brain. *Journal of Neurochemistry, 28*(6), 1221–1227.

Luisi, S., Petraglia, F., Benedetto, C., Nappi, R. E., Bernardi, F., Fadalti, M., … Genazzani, A. R. (2000). Serum allopregnanolone levels in pregnant women: Changes during pregnancy, at delivery, and in hypertensive patients. *Journal of Clinical Endocrinology & Metabolism, 85*(7), 2429–2433.

Macfarlane, S., Cleary, S., Bahrami, B., Reynolds, N., & Macfarlane, G. T. (2013). Synbiotic consumption changes the metabolism and composition of the gut microbiota in older people and modifies inflammatory processes: A randomised, double-blind, placebo-controlled crossover study. *Alimentary Pharmacology & Therapeutics, 38*(7), 804–816. https://doi.org/10.1111/apt.12453.

Magiakou, M. A., Mastorakos, G., Rabin, D., Dubbert, B., Gold, P. W., & Chrousos, G. P. (1996). Hypothalamic corticotropin-releasing hormone suppression during the postpartum period: Implications for the increase in psychiatric manifestations at this time. *Journal of Clinical Endocrinology & Metabolism, 81*(5), 1912–1917. https://doi.org/10.1210/jcem.81.5.8626857.

Maguire, J., & Mody, I. (2008). GABA(A)R plasticity during pregnancy: Relevance to postpartum depression. *Neuron, 59*(2), 207–213. https://doi.org/10.1016/j.neuron.2008.06.019.

Mahomed, K., Gülmezoglu, A. M., Nikodem, V. C., Wolman, W. L., Chalmers, B. E., & Hofmeyr, G. J. (1995). Labor experience, maternal mood and cortisol and catecholamine levels in low-risk primiparous women. *Journal of Psychosomatic Obstetrics and Gynaecology, 16*(4), 181–186.

Mastorakos, G., & Ilias, I. (2003). Maternal and fetal hypothalamic-pituitary-adrenal axes during pregnancy and postpartum. *Annals of the New York Academy of Sciences, 997*, 136–149.

McCoy, S. J. B., Beal, J. M., Payton, M. E., Stewart, A. L., DeMers, A. M., & Watson, G. H. (2008). Postpartum thyroid measures and depressive symptomology: A pilot study. *The Journal of the American Osteopathic Association, 108*(9), 503–507.

McQueen, J. K., Wilson, H., & Fink, G. (1997). Estradiol-17 beta increases serotonin transporter (SERT) mRNA levels and the density of SERT-binding sites in female rat brain. *Molecular Brain Research, 45*(1), 13–23.

Meltzer-Brody, S., Colquhoun, H., Riesenberg, R., Epperson, C. N., Deligiannidis, K. M., Rubinow, D. R., … Kanes, S. (2018). Brexanolone injection in post-partum depression: Two multicentre, double-blind, randomised, placebo-controlled, phase 3 trials. *Lancet (London, England)*. https://doi.org/10.1016/S0140-6736(18)31551-4.

Michopoulos, V., Berga, S. L., & Wilson, M. E. (2011). Estradiol and progesterone modify the effects of the serotonin reuptake transporter polymorphism on serotonergic responsivity to citalopram. *Experimental and Clinical Psychopharmacology, 19*(6), 401–408. https://doi.org/10.1037/a0025008.

Molendijk, M. L., Spinhoven, P., Polak, M., Bus, B. a. A., Penninx, B. W. J. H., & Elzinga, B. M. (2014). Serum BDNF concentrations as peripheral manifestations of depression: Evidence from a systematic review and meta-analyses on 179 associations (N=9484). *Molecular Psychiatry, 19*(7), 791–800. https://doi.org/10.1038/mp.2013.105.

Moleti, M., Trimarchi, F., & Vermiglio, F. (2014). Thyroid physiology in pregnancy. *Endocrine Practice: Official Journal of the American College of Endocrinology and the American Association of Clinical Endocrinologists, 20*(6), 589–596. https://doi.org/10.4158/EP13341.RA.

Moreno, F. A., McGahuey, C. A., Freeman, M. P., & Delgado, P. L. (2006). Sex differences in depressive response during monoamine depletions in remitted depressive subjects. *The Journal of Clinical Psychiatry, 67*(10), 1618–1623.

Moses-Kolko, E. L., Fraser, D., Wisner, K. L., James, J. A., Saul, A. T., Fiez, J. A., & Phillips, M. L. (2011). Rapid habituation of ventral striatal response to reward receipt in postpartum depression. *Biological Psychiatry, 70*(4), 395–399. https://doi.org/10.1016/j.biopsych.2011.02.021.

Moses-Kolko, E. L., Wisner, K. L., Price, J. C., Berga, S. L., Drevets, W. C., Hanusa, B. H., ... Meltzer, C. C. (2008). Serotonin 1A receptor reductions in postpartum depression: A positron emission tomography study. *Fertility and Sterility, 89*(3), 685–692. https://doi.org/10.1016/j.fertnstert.2007.03.059.

Muzik, M., Umarji, R., Sexton, M. B., & Davis, M. T. (2017). Family social support modifies the relationships between childhood maltreatment severity, economic adversity and postpartum depressive symptoms. *Maternal and Child Health Journal, 21*(5), 1018–1025. https://doi.org/10.1007/s10995-016-2197-4.

Nappi, R. E., Petraglia, F., Luisi, S., Polatti, F., Farina, C., & Genazzani, A. R. (2001). Serum allopregnanolone in women with postpartum "blues". *Obstetrics & Gynecology, 97*(1), 77–80.

Naseribafrouei, A., Hestad, K., Avershina, E., Sekelja, M., Linløkken, A., Wilson, R., & Rudi, K. (2014). Correlation between the human fecal microbiota and depression. *Neuro-Gastroenterology and Motility: The Official Journal of the European Gastrointestinal Motility Society, 26*(8), 1155–1162. https://doi.org/10.1111/nmo.12378.

Nierop, A., Bratsikas, A., Zimmermann, R., & Ehlert, U. (2006). Are stress-induced cortisol changes during pregnancy associated with postpartum depressive symptoms? *Psychosomatic Medicine, 68*(6), 931–937. https://doi.org/10.1097/01.psy.0000244385.93141.3b.

Okun, M. L., Luther, J., Prather, A. A., Perel, J. M., Wisniewski, S., & Wisner, K. L. (2011). Changes in sleep quality, but not hormones predict time to postpartum depression recurrence. *Journal of Affective Disorders, 130*(3), 378–384. https://doi.org/10.1016/j.jad.2010.07.015

Osborne, L. M., Gispen, F., Sanyal, A., Yenokyan, G., Meilman, S., & Payne, J. L. (2017). Lower allopregnanolone during pregnancy predicts postpartum depression: An exploratory study. *Psychoneuroendocrinology, 79*, 116–121. https://doi.org/10.1016/j.psyneuen.2017.02.012.

O'Hara, M. W., & Wisner, K. L. (2014). Perinatal mental illness: Definition, description and aetiology. *Best Practice & Research Clinical Obstetrics & Gynaecology, 28*(1), 3–12. https://doi.org/10.1016/j.bpobgyn.2013.09.002.

O'Keane, V., Lightman, S., Patrick, K., Marsh, M., Papadopoulos, A. S., Pawlby, S., ... Moore, R. (2011). Changes in the maternal hypothalamic-pituitary-adrenal axis during the early puerperium may be related to the postpartum "blues". *Journal of Neuroendocrinology, 23*(11), 1149–1155. https://doi.org/10.1111/j.1365-2826.2011.02139.x.

Pariante, C. M., & Lightman, S. L. (2008). The HPA axis in major depression: Classical theories and new developments. *Trends in Neurosciences, 31*(9), 464–468. https://doi.org/10.1016/j.tins.2008.06.006.

Payne, J. L., Palmer, J. T., & Joffe, H. (2009). A reproductive subtype of depression: conceptualizing models and moving toward etiology. *Harvard Review of Psychiatry, 17*(2), 72–86. https://doi.org/10.1080/10673220902899706.

Pedersen, C., Leserman, J., Garcia, N., Stansbury, M., Meltzer-Brody, S., & Johnson, J. (2016). Late pregnancy thyroid-binding globulin predicts perinatal depression. *Psychoneuroendocrinology, 65*, 84–93. https://doi.org/10.1016/j.psyneuen.2015.12.010.

Raison, C. L., Capuron, L., & Miller, A. H. (2006). Cytokines sing the blues: Inflammation and the pathogenesis of depression. *Trends in Immunology, 27*(1), 24–31. https://doi.org/10.1016/j.it.2005.11.006.

Reigstad, C. S., Salmonson, C. E., Rainey, J. F., Szurszewski, J. H., Linden, D. R., Sonnenburg, J. L., ... Kashyap, P. C. (2015). Gut microbes promote colonic serotonin production through an effect of short-chain fatty acids on enterochromaffin cells. *Federation of American Societies for Experimental Biology Journal: Official Publication of the Federation of American Societies for Experimental Biology, 29*(4), 1395–1403. https://doi.org/10.1096/fj.14-259598.

Rossetti, M. F., Cambiasso, M. J., Holschbach, M. A., & Cabrera, R. (2016). Oestrogens and progestagens: Synthesis and action in the brain. *Journal of Neuroendocrinology, 28*(7). https://doi.org/10.1111/jne.12402.

Röytiö, H., Mokkala, K., Vahlberg, T., & Laitinen, K. (2017). Dietary intake of fat and fibre according to reference values relates to higher gut microbiota richness in overweight pregnant women. *The British Journal of Nutrition, 118*(5), 343–352. https://doi.org/10.1017/S0007114517002100.

Sacher, J., Wilson, A. A., Houle, S., Rusjan, P., Hassan, S., Bloomfield, P. M., ... Meyer, J. H. (2010). Elevated brain monoamine oxidase A binding in the early postpartum period. *Archives of General Psychiatry, 67*(5), 468–474. https://doi.org/10.1001/archgenpsychiatry.2010.32.

Scheyer, K., & Urizar, G. G. (2016). Altered stress patterns and increased risk for postpartum depression among low-income pregnant women. *Archives of Women's Mental Health, 19*(2), 317–328. https://doi.org/10.1007/s00737-015-0563-7.

Schiller, C. E., Meltzer-Brody, S., & Rubinow, D. R. (2015). The role of reproductive hormones in postpartum depression. *CNS Spectrums, 20*(1), 48–59. https://doi.org/10.1017/S1092852914000480.

Schock, H., Zeleniuch-Jacquotte, A., Lundin, E., Grankvist, K., Lakso, H.-Å., Idahl, A., ... Fortner, R. T. (2016). Hormone concentrations throughout uncomplicated pregnancies: A longitudinal study. *BMC Pregnancy and Childbirth, 16*(1), 146. https://doi.org/10.1186/s12884-016-0937-5.

Sell, S. L., Craft, R. M., Seitz, P. K., Stutz, S. J., Cunningham, K. A., & Thomas, M. L. (2008). Estradiol-sertraline synergy in ovariectomized rats. *Psychoneuroendocrinology, 33*(8), 1051–1060. https://doi.org/10.1016/j.psyneuen.2008.05.006.

Seth, S., Lewis, A. J., & Galbally, M. (2016). Perinatal maternal depression and cortisol function in pregnancy and the postpartum period: A systematic literature review. *BMC Pregnancy and Childbirth, 16*. https://doi.org/10.1186/s12884-016-0915-y.

Shanks, N., & Lightman, S. L. (2001). The maternal-neonatal neuro-immune interface: Are there long-term implications for inflammatory or stress-related disease? *Journal of Clinical Investigation, 108*(11), 1567–1573.

Sichel, D. A., Cohen, L. S., Robertson, L. M., Ruttenberg, A., & Rosenbaum, J. F. (1995). Prophylactic estrogen in recurrent postpartum affective disorder. *Biological Psychiatry, 38*(12), 814–818. https://doi.org/10.1016/0006-3223(95)00063-1.

Slykerman, R. F., Hood, F., Wickens, K., Thompson, J. M. D., Barthow, C., Murphy, R., ... Probiotic in Pregnancy Study Group. (2017). Effect of *Lactobacillus rhamnosus* HN001 in pregnancy on postpartum symptoms of depression and anxiety: A randomised double-blind placebo-controlled trial. *EBioMedicine, 24*, 159–165. https://doi.org/10.1016/j.ebiom.2017.09.013.

Smith, S. S., Shen, H., Gong, Q. H., & Zhou, X. (2007). Neurosteroid regulation of GABA(A) receptors: Focus on the alpha4 and delta subunits. *Pharmacology & Therapeutics, 116*(1), 58–76. https://doi.org/10.1016/j.pharmthera.2007.03.008.

Stuebe, A. M., Grewen, K., & Meltzer-Brody, S. (2013). Association between maternal mood and oxytocin response to breastfeeding. *Journal of Women's Health, 22*(4), 352–361. https://doi.org/10.1089/jwh.2012.3768.

Sylvén, S. M., Elenis, E., Michelakos, T., Larsson, A., Olovsson, M., Poromaa, I. S., & Skalkidou, A. (2013). Thyroid function tests at delivery and risk for postpartum depressive symptoms. *Psychoneuroendocrinology, 38*(7), 1007–1013. https://doi.org/10.1016/j.psyneuen.2012.10.004.

Szpunar, M. J., & Parry, B. L. (2018). A systematic review of cortisol, thyroid-stimulating hormone, and prolactin in peripartum women with major depression. *Archives of Women's Mental Health, 21*(2), 149–161. https://doi.org/10.1007/s00737-017-0787-9.

Taylor, A., Glover, V., Marks, M., & Kammerer, M. (2009). Diurnal pattern of cortisol output in postnatal depression. *Psychoneuroendocrinology, 34*(8), 1184–1188. https://doi.org/10.1016/j.psyneuen.2009.03.004.

Timby, E., Bäckström, T., Nyberg, S., Stenlund, H., Wihlbäck, A.-C. N., & Bixo, M. (2016). Women with premenstrual dysphoric disorder have altered sensitivity to allopregnanolone over the menstrual cycle compared to controls-a pilot study. *Psychopharmacology*. https://doi.org/10.1007/s00213-016-4258-1.

Togher, K. L., Khashan, A. S., Kenny, L. C., Stanton, C., Carafa, I., Murphy, K., … Clarke, G. (2017). 42 depression-associated alterations in the maternal microbiome during pregnancy: Priming for adverse infant outcomes? *Gut, 66*(Suppl. 1). A15–A15 https://doi.org/10.1136/gutjnl-2017-314127.42.

Toufexis, D. J., Davis, C., Hammond, A., & Davis, M. (2004). Progesterone attenuates corticotropin-releasing factor-enhanced but not fear-potentiated startle via the activity of its neuroactive metabolite, allopregnanolone. *The Journal of Neuroscience: The Official Journal of the Society for Neuroscience, 24*(45), 10280–10287. https://doi.org/10.1523/JNEUROSCI.1386-04.2004.

Vannuccini, S., Bocchi, C., Severi, F. M., Challis, J. R., & Petraglia, F. (2016). Endocrinology of human parturition. *Annales d'Endocrinologie,* 77(2), 105–113. https://doi.org/10.1016/j.ando.2016.04.025.

Veen, C., Myint, A. M., Burgerhout, K. M., Schwarz, M. J., Schütze, G., Kushner, S. A., … Bergink, V. (2016). Tryptophan pathway alterations in the postpartum period and in acute postpartum psychosis and depression. *Journal of Affective Disorders, 189*, 298–305. https://doi.org/10.1016/j.jad.2015.09.064.

Wang, Y., Telesford, K. M., Ochoa-Repáraz, J., Haque-Begum, S., Christy, M., Kasper, E. J., … Kasper, L. H. (2014). An intestinal commensal symbiosis factor controls neuroinflammation via TLR2-mediated CD39 signalling. *Nature Communications, 5*, 4432. https://doi.org/10.1038/ncomms5432.

Wissart, J., Parshad, O., & Kulkarni, S. (2005). Prevalence of pre- and postpartum depression in Jamaican women. *BMC Pregnancy and Childbirth, 5*, 15. https://doi.org/10.1186/1471-2393-5-15.

Xie, R., Xie, H., Krewski, D., & He, G. (2018). Plasma concentrations of neurotransmitters and postpartum depression. *Zhong Nan Da Xue Xue Bao. Yi Xue Ban (Journal of Central South University. Medical Sciences), 43*(3), 274–281. https://doi.org/10.11817/j.issn.1672-7347.2018.03.007.

Yildiz, G., Senturk, M. B., Yildiz, P., Cakmak, Y., Budak, M. S., & Cakar, E. (2017). Serum serotonin, leptin, and adiponectin changes in women with postpartum depression: Controlled study. *Archives of Gynecology and Obstetrics, 295*(4), 853–858. https://doi.org/10.1007/s00404-017-4313-0.

Yim, I. S., Glynn, L. M., Dunkel-Schetter, C., Hobel, C. J., Chicz-DeMet, A., & Sandman, C. A. (2009). Risk of postpartum depressive symptoms with elevated corticotropin-releasing hormone in human pregnancy. *Archives of General Psychiatry, 66*(2), 162–169. https://doi.org/10.1001/archgenpsychiatry.2008.533.

Yim, I. S., Tanner Stapleton, L. R., Guardino, C. M., Hahn-Holbrook, J., & Schetter, C. D. (2015). Biological and psychosocial predictors of postpartum depression: Systematic review and call for integration. *Annual Review of Clinical Psychology, 11*, 99–137. https://doi.org/10.1146/annurev-clinpsy-101414-020426.

Yunes, R. A., Poluektova, E. U., Dyachkova, M. S., Klimina, K. M., Kovtun, A. S., Averina, O. V., … Danilenko, V. N. (2016). GABA production and structure of gadB/gadC genes in *Lactobacillus* and *Bifidobacterium* strains from human microbiota. *Anaerobe, 42*, 197–204. https://doi.org/10.1016/j.anaerobe.2016.10.011.

CHAPTER 5

Immunological biomarkers of postpartum depression

Lauren M. Osborne[1,2,3]

[1]Department of Psychiatry & Behavioral Sciences, The Johns Hopkins University School of Medicine, Baltimore, MD, United States of America; [2]Department of Gynecology & Obstetrics, The Johns Hopkins University School of Medicine, Baltimore, MD, United States of America; [3]Women's Mood Disorders Center, The Johns Hopkins University School of Medicine, Baltimore, MD, United States of America

Introduction

The connection between major depressive disorder (MDD) and immune dysregulation is at this point well established. Numerous studies show evidence of elevations of proinflammatory cytokines in some individuals with depression (Dowlati et al., 2010; Howren, Lamkin, & Suls, 2009; Smith, 1991); other research has clarified that immune dysregulation may play a role in depression for only certain subsets of people with depression (for example, those with elevated inflammatory markers at baseline) (Miller, Maletic, & Raison, 2009). Inflammation is not the only connection between the immune system and depression; numerous factors concerning cell-mediated immunity and nitrosative and oxidative stress may also be related to the onset of depressive symptoms (Leonard and Maes, 2012). Gaining additional insights into the role of immune dysregulation in depression may require investigations into particular subsets of people with depression. Postpartum women are one such subset in whom there are obvious and significant changes in the immune system. Shedding light on immune dysregulation in postpartum depression (PPD) may not only help women with PPD, but may also help us to understand the pathophysiology of immune related depression more generally. While the literature on immune dysregulation is still small, we now have a body of evidence that, taken together, may help us to understand promising directions in the search for biomarkers of PPD.

The immune system, the body's defense against foreign pathogens, undergoes considerable changes across the perinatal period. The fetus is a semiallogeneic presence, necessitating serious alterations in the mother's immune system, both to prevent rejection of the fetus and to prevent

Biomarkers of Postpartum Psychiatric Disorders
ISBN 978-0-12-815508-0
https://doi.org/10.1016/B978-0-12-815508-0.00005-9

further hypersensitivity reactions on the part of the mother. Early research on perinatal immune modulation assumed pregnancy to be a time of immune suppression, but more recent research supports a more complex model, in which innate immune barriers are strengthened, but some aspects of adaptive immunity are less effective (Chen, Liu, & Sytwu, 2012; Holtan et al., 2015; Kraus et al., 2010, 2012; Pazos, Sperling, Moran, & Kraus, 2012), with especially significant changes in the T cell—mediated immune response. Women in late pregnancy are especially vulnerable to pathogens, with increased rates of mortality from both bacterial and viral infections (Sherer, Posillico, & Schwarz, 2018). In addition, autoimmune disease expression is altered during pregnancy, with some autoimmune diseases (e.g., rheumatoid arthritis and multiple sclerosis) going into relative remission in pregnancy while others (e.g., systemic lupus erythematosus) flare during pregnancy; asthma and atopic diseases are variable (Fischer-Betz & Specker, 2017; Grosso et al., 2018; Voskuhl & Momtazee, 2017). (For greater detail on the psychoneuroimmunology of pregnancy, see recent reviews [Leff-Gelman et al., 2016; Sherer et al., 2018].)

The substantial immune changes of pregnancy must then be further modulated in the postpartum, when both mother and baby are vulnerable to infection and the baby continues to rely on the maternal immune system through breast milk (Le Doare, Holder, Bassett, & Pannaraj, 2018) for protection. Some aspects of the adaptive immune system appear to rebound almost immediately in the normal postpartum woman, including T cell responses (Bergink et al., 2013), the fever response (Martin, Malkinson, Veale, & Pittman, 1995), and autoimmune activity (Confavreux, Hutchinson, Hours, Cortinovis-Tourniaire, & Moreau, 1998; Hazes et al., 2011; Jalkanen, Saraste, Gfeller, Surcel, & Airas, 2013; Wesseloo, Kamperman, Bergink, & Pop, 2018). There is still much to be done to determine the exact timetable of immune system rebound in the postpartum, with some evidence showing almost immediate changes after parturition (Burns et al., 1999), but evidence from some longitudinal studies showing that changes occurring in pregnancy can persist until at least 6 months postpartum (Osborne, Yenokyan, et al., 2018).

An early review of inflammation and perinatal depression (Osborne & Monk, 2013) highlighted the burgeoning interest in examining perinatal depression as an inflammatory morbidity of pregnancy. The evidence cited in that review covered both antenatal and postpartum depression, with some studies conflating the two. It was clear that, while substantial evidence pointed to correlations between immune dysregulation and perinatal depression, a lack of consistent research targets and consistent tools of measurement meant that it was too soon to draw conclusions about the directionality of this association—and far too soon to ask whether there were immune targets that could be considered biomarkers of disease in the perinatal period.

In this chapter, I hope to augment that prior review—focusing here on PPD alone—by summarizing those results and updating them with evidence from the last 5 years. I will use the term "PPD" throughout, while acknowledging that many studies are actually measuring postpartum depressive symptoms by scores on a scale, rather than syndromic PPD as evaluated by a clinician using DSM criteria. I hope to synthesize this evidence with an eye toward possible immunological biomarkers of PPD—in other words, with what we now know about the role of immune dysregulation and PPD, are there promising immune targets that might, if measured in late pregnancy or early postpartum, point us toward women at high risk of developing PPD? And if so, are these targets modifiable, and what research remains to be done to develop novel therapeutics for these targets?

Early work

The earliest studies on immune changes in PPD were done in the 1990s. The first such study, in a prescient move, examined immune cell populations (an approach that we then ignored for the next 20 years, but is now considered to be a necessary next step to advance this field). Dysphoria (a variable combining results of the Stein Blues Scale, the Edinburgh Postnatal Depression Scale [EPDS], and the Hospital Anxiety and Depression Scale) was negatively correlated with total lymphocyte count in 75 postpartum women, a result that was found to be driven by T cell count—in other words, dysphoric postpartum women had lower T cell counts than healthy postpartum women (Hucklebridge et al., 1994).

Interleukin 6

Following this study, nearly all examinations of immune dysregulation and PPD have examined peripheral cytokines. The most commonly examined cytokine has been IL-6, a cytokine with both pro- and antiinflammatory properties that is integral to the acute phase response and to B and T cell activation. Women whose scores on the Zung Depression Rating Scale

(ZDS) increased in the postpartum had elevated levels of both IL-6 and its receptor at 1 day postpartum (Maes et al., 2000). In addition, women who had a history of past depression had significantly higher levels of serum IL-6 and its soluble receptor than women without such a history, though their ZDS scores were not different (Maes, Ombelet, De Jongh, Kenis, & Bosmans, 2001). Elevated serum IL-6 at delivery has predicted the later development of PPD (Liu, Zhang, Gao, & Zhang, 2016). Urinary IL-6 has been found to fall over the first 4 weeks postpartum in women without depression, but not in women with depression (though no correlation was found with depressive symptom scores and IL-6 at 4 weeks postpartum in this study) (Corwin, Johnston, & Pugh, 2008). IL-6 as measured in the cerebrospinal fluid (CSF) has also been found to be correlated with EPDS scores at postpartum days 1–4 and at postpartum weeks 1 and 6 (though this same study did not fine a correlation for serum IL-6) (Boufidou et al., 2009). In preterm mothers, both serum and cord blood IL-6 have been positively correlated with negative affect (Fransson et al., 2012). IL-6 measured in the third trimester was a significant predictor of postpartum EPDS scores (Simpson, Steiner, Coote, & Frey, 2016). Corwin et al. (2015) found no relationship between IL-6 and depressive symptoms when measured in the third trimester or early postpartum, but found that cortisol AUC increased with increasing levels of IL-6 for women with depressive symptoms but not for those without at 14 days postpartum. Animal literature has also examined serum IL-6, which is elevated in one rat model of PPD (Li et al., 2016). Not all studies have found a positive association, however; in one of the largest and most rigorous studies (in 347 women), there was no association between serum IL-6 at delivery and EPDS scores at multiple subsequent points in the postpartum (Skalkidou et al., 2009), and another study of 56 women with a past history of MDD or PPD found no correlation between depression recurrence and IL-6 levels at an unspecified point in the first 17 weeks postpartum (Okun et al., 2011).

Other cytokines

Purely proinflammatory cytokines (and their receptors) have been examined as well, with mixed results. In 181 healthy mothers at 4–6 weeks postpartum, increased mood disturbance (as measured by the Profile of Mood States and a Total Mood Disturbance Score [TMDS] calculated by the researchers) was significantly negatively correlated with levels of IFN-γ and with the IFN-γ/IL-10 ratio (Groer & Davis, 2006). IFN-γ is a characteristic proinflammatory cytokine produced by $CD4^+$ Th1 cells that activates macrophages and is crucial to the body's response to intravesicular pathogens. This correlation was found in formula-feeding mothers only. In addition, when women were divided into deciles according to depression status, those in the highest decile had significantly lower levels of IFN-γ (Groer & Morgan, 2007). TNF-α measured in the CSF at delivery has been positively associated with EPDS score at 1 week postpartum (Boufidou et al., 2009). The complicated marker TGF-β, which seems to play a dual role, has also been examined. Decreased TGF-β is thought to render the individual more susceptible to immune dysregulation and the rise of a proinflammatory state, a dysregulation that can then lead to *increase*d levels of TGF-β, which together with IL-6 can lead to increases in Th17 cells (Saxena et al., 2008). Depressed mothers have been found to have a 4.4-fold (CI 1.5–13) higher likelihood of having increased levels of TGF-β in their breast milk (Kondo et al., 2011). Serum IL-8, an important mediator of the innate immune response whose primary purpose is the induction of chemotaxis in neutrophils and other target cells, has also been associated with depressive symptoms in the first 5 days postpartum in women who delivered preterm (with no such association found for full-term mothers) (Fransson et al., 2012). In the same study, again in pre-term women, both serum and cord blood IL-18 (a proinflammatory cytokine produced by macrophages and epithelial cells that induces the production of IFN-g) were associated with depressive symptoms and negative affect (Fransson et al., 2012). Finally, serum IL-1β, an important mediator of the inflammatory and pyrogenic response that is produced by activated macrophages, has been elevated in one rat model of PPD (Li et al., 2016).

The bulk of these studies have found elevations in proinflammatory cytokines to be associated with PPD. The flip side of this is decreases in antiinflammatory cytokines, and a smaller number of studies have examined either antiinflammatory cytokine levels or the ratio between pro- and antiinflammatory (using two characteristic cytokines as a marker of this ratio). Women later diagnosed with PPD had lower levels of an antiinflammatory protein (Clara Cell Protein 16, CC16) both 3–5 days before and 1–3 days after giving birth. In this study, PPD was defined as an onset of depression within 3 months of parturition but was assessed by a structured clinical interview administered 6–10 months later (Maes et al., 1999). Some studies, however, have instead found relative increases in antiinflammatory action, which some speculate may be compensatory. The ratio of IFN-γ to IL-10 was negatively correlated with increased scores on the TMDS at 4–6 weeks postpartum (Groer & Davis, 2006) (indicating either decreases in IFN-γ, increases in IL-10, or both). IL-1ra, a protein that binds to the IL-1 receptor and inhibits the proinflammatory effect of IL-1β, was elevated in postpartum women with a history of past depression (though the women did not differ in terms of concurrent depressive scores) (Maes et al., 2001).

More complex immune markers and associations

The most recent human studies have examined a broader set of immune markers and looked at complex relationships among them and between the immune system and other systems. A recent Swedish study examined over 70 inflammatory measures in the postpartum and found elevated levels of five of these (TRANCE, HGF, IL-18, FGF 23, and CXCL-1) in women with PPD (Brann et al., 2018). These markers were examined concurrently with symptoms, so at this point cannot be useful as predictive biomarkers. A second study by the same group did attempt to offer a predictive conclusion; this study examined 92 inflammatory markers in late pregnancy and sought to determine whether they were related to depressive symptoms at 8 weeks postpartum. Five of these (STAM-BP [or otherwise AMSH], AXIN-1, ADA, ST1A1, and IL-10) were decreased in late pregnancy in women who went on to have postpartum depressive symptoms. In addition, the authors tested a summary inflammation variable, which predicted the development of PPD (though not as well as a prior history of depression) (Brann et al., 2017). Another group looked at relationships among stress, fatigue, depressive symptoms, and cytokines both pre- and postpartum, finding that IL-5 was significantly associated with postpartum depressive symptoms—but they did not examine whether immune functioning in pregnancy was related to later postpartum depressive symptoms (Cheng & Pickler, 2014). Another broad group of immune markers in pregnancy was examined specifically for its ability to predict postpartum depressive symptoms in 100 pregnant women. Diverse markers including neopterin, regulatory T cells, CXCR1, CCR2, MNP1, and CD11a were measured by flow cytometry or ELISA, and women with depressive symptoms had prenatal elevations in T regulatory cells and neopterin as well as a greater decrease in CXCR1 from pregnancy to the postpartum than did healthy women (Krause et al., 2014). My own exploratory study has found that elevations in several proinflammatory cytokines are associated with depressive symptoms in late pregnancy, but the population had few women with postpartum depressive symptoms, so it is unclear if these elevations would predict the postpartum depressive state (Osborne, Yenokyan, et al., 2018). Other studies, however, have found no association between postpartum immune functioning and depressive symptoms (Groer, Jevitt, & Ji, 2015).

One study looked not at markers of inflammation but at antibodies, finding that the presence of thyroid peroxidase antibodies early in gestation in a cohort of over a 1000 women increased the odds of developing first-onset PPD almost fourfold (adjusted OR 3.8; 95% CI 1.3–11.6) when symptoms were measured at 4 months postpartum (but not at 6 weeks or 8 or 12 months postpartum) (Wesseloo et al., 2018).

Of course, the immune system does not function in isolation, and one interesting study looked at the interplay between the immune system and the HPA axis. This study measured markers of both systems in late pregnancy and at multiple time points in the postpartum and found that cortisol and the IL8/IL10 ratio at day 14, considered together, significantly predicted the development of PPD symptoms as measured by the EPDS. They also found that levels of TNF-α were significantly lower in the postpartum in women with depressive symptoms, and they hypothesized that it was not the levels of cytokines themselves but the response of the HPA axis to cytokines that was the causative factor. In other words, an increase in proinflammatory cytokines is normal for the postpartum period, but the HPA axis response to that is ordinarily blunted—whereas women with PPD symptoms instead had a hyperactive cortisol response to the increased immune activity. When family history of depression was included in the model, it predicted PPD with an accuracy of about 84% (Corwin et al., 2015).

While not associated with a particular measurable immunological biomarker, we should also make note of the literature connecting an immunological morbidity of pregnancy (preeclampsia) with subsequent PPD. A Danish registry study found that primiparous preeclamptic women had a risk of PPD elevated fourfold above that of the general population (IRR 4.21, 95% CI 2.89–6.13) (Bergink et al., 2015). Other articles have pointed to a link between PPD and dysregulation of tryptophan metabolism, which is altered by inflammatory activity (Duan et al., 2018) and has an impact on the production of serotonin.

Animal literature has also contributed in recent years and is able to offer more direct measurements of immune activity in the brain. There is a significant reduction in microglial density in late pregnancy and early postpartum in healthy rats (Haim et al., 2017). IL-1β, measured directly in brain tissue and microglia, was decreased in the postpartum in rats that underwent a forced swimming stressor in late pregnancy. The same study showed that this pregnancy stressor could induce depressive-like behaviors in the postpartum, and that antenatal immune activation predicted an attenuated immune response (specifically in IL-1β and CD11b) when cells from dams with postpartum depressive-like behaviors were artificially stimulated in the laboratory to produce an inflammatory response. These results indicate that postpartum anhedonia in this sample may reflect an antiinflammatory state (as opposed to the proinflammatory state associated with depression outside the postpartum period, and in contrast to some of the human literature cited above) (Posillico & Schwarz, 2016).

Discussion

What are we to make of the literature to date, and can it help us in a quest for immunological biomarkers of PPD? While the number of studies has increased in recent years, the conclusions one can draw remain opaque. Certainly, there is considerable evidence of a role for IL-6, with the majority of studies showing elevation in PPD and some studies showing its predictive value when measured in pregnancy. Not all studies reach the same conclusions, however, and IL-6 is a complicated cytokine with both pro- and antiinflammatory properties. Evidence is mixed on other proinflammatory markers, with some studies showing increases and others showing decreases. Timing may be a crucial element here, with the immediate postpartum period being a different immune environment than later in the postpartum. In addition, timing of the measurement of postpartum depressive symptoms may also be crucial. Many women have mood disturbance in the first few days postpartum that does not go on to result in PPD. Genetic evidence indicates that depression arising in the first 4 weeks may be biologically distinct from depression later in the postpartum period (Forty et al., 2006; Payne, Palmer, & Joffe, 2009). Studies that examine antiinflammatory activity are too few to draw any conclusions. In addition, most studies have considered peripheral cytokines, which can be altered by any number of circumstances, including racial and ethnic differences, and by factors that differ between depressed and nondepressed populations, such as obesity and trauma history (Christian, Kowalsky, Mitchell, & Porter, 2018; Finy & Christian, 2018; Mitchell, Porter, & Christian, 2018).

If the measurement of individual peripheral cytokines is not likely to yield a useful immunological biomarker of PPD, is there another possible direction for the future of immunological research in PPD? Two promising avenues have been indicated in recent studies. The first is the concept of a model including multiple elements—either as a summary inflammation score (Brann et al., 2017) or as a model including other systems, such as the HPA axis, in a summary prediction model (Corwin et al., 2015). These approaches, while increasing the positive predictive value of tests for PPD, do not yet seem to be clinically useful, as the burden on patients for a test that involves multiple elements (blood and saliva measurements, for example) is too high to render the tests more valuable than current psychosocial and family history predictors that can be determined merely by asking the patient. A test that could use one blood test to combine different systems—something akin to the concept of allostatic load (Maestripieri & Hoffman, 2011)—may hold promise.

Another direction that may yield possible biomarkers is measurement of more complex elements of the immune system—for example, T cell counts or other leukocyte measurements. While any test requiring measurement of the production of stimulated immune cells is at present too cumbersome and expensive for daily use, it is possible that measurements of peripheral immune cell types may yield important information that could be used in a predictive test. One study has already shown that mothers with PPD fail to show the normal postpartum rebound in Th1 cells, and increasing evidence about the many and varied roles of Th17 cells may yield useful directions (Osborne, Brar, & Klein, 2018).

One thing is certain—while increased research into the immune system will enhance our understanding of the pathophysiology of PPD, there is no biomarker on the horizon that will increase the predictive value we can currently derive from appropriately screening patients for family and personal history and for psychosocial risk factors. Until we can consistently implement these basic and low-cost techniques to improve our ability to detect PPD, the search for more complicated biological predictors may not be our highest priority.

References

Bergink, V., Burgerhout, K. M., Weigelt, K., Pop, V. J., de Wit, H., Drexhage, R. C., ... Drexhage, H. A. (2013). Immune system dysregulation in first-onset postpartum psychosis. *Biological Psychiatry, 73*(10), 1000–1007. https://doi.org/10.1016/j.biopsych.2012.11.006.

Bergink, V., Laursen, T. M., Johannsen, B. M., Kushner, S. A., Meltzer-Brody, S., & Munk-Olsen, T. (2015). Pre-eclampsia and first-onset postpartum psychiatric episodes: A Danish population-based cohort study. *Psychological Medicine, 45*(16), 3481–3489. https://doi.org/10.1017/s0033291715001385.

Boufidou, F., Lambrinoudaki, I., Argeitis, J., Zervas, I. M., Pliatsika, P., Leonardou, A. A., … Nikolaou, C. (2009). CSF and plasma cytokines at delivery and postpartum mood disturbances. *Journal of Affective Disorders, 115*(1–2), 287–292. https://doi.org/10.1016/j.jad.2008.07.008.

Brann, E., Fransson, E., White, R. A., Papadopoulos, F. C., Edvinsson, A., Kamali-Moghaddam, M., … Skalkidou, A. (2018). Inflammatory markers in women with postpartum depressive symptoms. *Journal of Neuroscience Research*. https://doi.org/10.1002/jnr.24312.

Brann, E., Papadopoulos, F., Fransson, E., White, R., Edvinsson, A., Hellgren, C., … Skalkidou, A. (2017). Inflammatory markers in late pregnancy in association with postpartum depression—A nested case-control study. *Psychoneuroendocrinology, 79*, 146–159. https://doi.org/10.1016/j.psyneuen.2017.02.029.

Burns, D. N., Nourjah, P., Wright, D. J., Minkoff, H., Landesman, S., Rubinstein, A., … Nugent, R. P. (1999). Changes in immune activation markers during pregnancy and postpartum. *Journal of Reproductive Immunology, 42*(2), 147–165.

Chen, S. J., Liu, Y. L., & Sytwu, H. K. (2012). Immunologic regulation in pregnancy: From mechanism to therapeutic strategy for immunomodulation. *Clinical and Developmental Immunology, 2012*, 258391. https://doi.org/10.1155/2012/258391.

Cheng, C. Y., & Pickler, R. H. (2014). Perinatal stress, fatigue, depressive symptoms, and immune modulation in late pregnancy and one month postpartum. *The Scientific World Journal, 2014*, 652630. https://doi.org/10.1155/2014/652630.

Christian, L. M., Kowalsky, J. M., Mitchell, A. M., & Porter, K. (2018). Associations of postpartum sleep, stress, and depressive symptoms with LPS-stimulated cytokine production among African American and White women. *Journal of Neuroimmunology, 316*, 98–106. https://doi.org/10.1016/j.jneuroim.2017.12.020.

Confavreux, C., Hutchinson, M., Hours, M. M., Cortinovis-Tourniaire, P., & Moreau, T. (1998). Rate of pregnancy-related relapse in multiple sclerosis. pregnancy in multiple sclerosis group. *New England Journal of Medicine, 339*(5), 285–291. https://doi.org/10.1056/nejm199807303390501.

Corwin, E. J., Johnston, N., & Pugh, L. (2008). Symptoms of postpartum depression associated with elevated levels of interleukin-1 beta during the first month postpartum. *Biological Research For Nursing, 10*(2), 128–133. https://doi.org/10.1177/1099800408323220.

Corwin, E. J., Pajer, K., Paul, S., Lowe, N., Weber, M., & McCarthy, D. O. (2015). Bidirectional psychoneuroimmune interactions in the early postpartum period influence risk of postpartum depression. *Brain, Behavior, and Immunity, 49*, 86–93. https://doi.org/10.1016/j.bbi.2015.04.012.

Dowlati, Y., Herrmann, N., Swardfager, W., Liu, H., Sham, L., Reim, E. K., & Lanctot, K. L. (2010). A meta-analysis of cytokines in major depression. *Biological Psychiatry, 67*(5), 446–457. https://doi.org/10.1016/j.biopsych.2009.09.033.

Duan, K. M., Ma, J. H., Wang, S. Y., Huang, Z., Zhou, Y., & Yu, H. (2018). The role of tryptophan metabolism in postpartum depression. *Metabolic Brain Disease, 33*(3), 647–660. https://doi.org/10.1007/s11011-017-0178-y.

Finy, M. S., & Christian, L. M. (2018). Pathways linking childhood abuse history and current socioeconomic status to inflammation during pregnancy. *Brain, Behavior, and Immunity, 74*, 231–240. https://doi.org/10.1016/j.bbi.2018.09.012.

Fischer-Betz, R., & Specker, C. (2017). Pregnancy in systemic lupus erythematosus and antiphospholipid syndrome. *Best Practice & Research Clinical Rheumatology, 31*(3), 397–414. https://doi.org/10.1016/j.berh.2017.09.011.

Forty, L., Jones, L., Macgregor, S., Caesar, S., Cooper, C., Hough, A., ... Jones, I. (2006). Familiality of postpartum depression in unipolar disorder: Results of a family study. *American Journal of Psychiatry, 163*(9), 1549–1553. https://doi.org/10.1176/ajp.2006.163.9.1549.

Fransson, E., Dubicke, A., Bystrom, B., Ekman-Ordeberg, G., Hjelmstedt, A., & Lekander, M. (2012). Negative emotions and cytokines in maternal and cord serum at preterm birth. *American Journal of Reproductive Immunology, 67*(6), 506–514. https://doi.org/10.1111/j.1600-0897.2011.01081.x.

Groer, M. W., & Davis, M. W. (2006). Cytokines, infections, stress, and dysphoric moods in breastfeeders and formula feeders. *Journal of Obstetric, Gynecologic, and Neonatal Nursing, 35*(5), 599–607. https://doi.org/10.1111/j.1552-6909.2006.00083.x.

Groer, M. E., Jevitt, C., & Ji, M. (2015). Immune changes and dysphoric moods across the postpartum. *American Journal of Reproductive Immunology, 73*(3), 193–198. https://doi.org/10.1111/aji.12322.

Groer, M. W., & Morgan, K. (2007). Immune, health and endocrine characteristics of depressed postpartum mothers. *Psychoneuroendocrinology, 32*(2), 133–139. https://doi.org/10.1016/j.psyneuen.2006.11.007.

Grosso, A., Locatelli, F., Gini, E., Albicini, F., Tirelli, C., Corveri, I., & Corsico, A. G. (2018). The course of asthma during pregnancy in a recent, multicase-control study on respiratory health. *Allergy, Asthma and Clinical Immunology, 14*, 16. https://doi.org/10.1186/s13223-018-0242-0.

Haim, A., Julian, D., Albin-Brooks, C., Brothers, H. M., Lenz, K. M., & Leuner, B. (2017). A survey of neuroimmune changes in pregnant and postpartum female rats. *Brain, Behavior, and Immunity, 59*, 67–78. https://doi.org/10.1016/j.bbi.2016.09.026.

Hazes, J. M., Coulie, P. G., Geenen, V., Vermeire, S., Carbonnel, F., Louis, E., ... De Keyser, F. (2011). Rheumatoid arthritis and pregnancy: Evolution of disease activity and pathophysiological considerations for drug use. *Rheumatology (Oxford), 50*(11), 1955–1968. https://doi.org/10.1093/rheumatology/ker302.

Holtan, S. G., Chen, Y., Kaimal, R., Creedon, D. J., Enninga, E. A., Nevala, W. K., & Markovic, S. N. (2015). Growth modeling of the maternal cytokine milieu throughout normal pregnancy: Macrophage-derived chemokine decreases as inflammation/counterregulation increases. *Journal of Immunology Research, 2015*. https://doi.org/10.1155/2015/952571, 952571.

Howren, M. B., Lamkin, D. M., & Suls, J. (2009). Associations of depression with C-reactive protein, IL-1, and IL-6: A meta-analysis. *Psychosomatic Medicine, 71*(2), 171–186. https://doi.org/10.1097/PSY.0b013e3181907c1b.

Hucklebridge, F. H., Smith, M. D., Clow, A., Evans, P., Glover, V., Taylor, A., ... Lydyard, P. M. (1994). Dysphoria and immune status in postpartum women. *Biological Psychology, 37*(3), 199–206.

Jalkanen, A., Saraste, M., Gfeller, A., Surcel, H. M., & Airas, L. (2013). Increased thyroid autoimmunity among women with multiple sclerosis in the postpartum setting. *Multiple Sclerosis, 19*(13), 1734–1742. https://doi.org/10.1177/1352458513485148.

Kondo, N., Suda, Y., Nakao, A., Oh-Oka, K., Suzuki, K., Ishimaru, K., ... Yamagata, Z. (2011). Maternal psychosocial factors determining the concentrations of transforming growth factor-beta in breast milk. *Pediatric Allergy & Immunology, 22*(8), 853–861. https://doi.org/10.1111/j.1399-3038.2011.01194.x.

Krause, D., Jobst, A., Kirchberg, F., Kieper, S., Hartl, K., Kastner, R., ... Schwarz, M. J. (2014). Prenatal immunologic predictors of postpartum depressive symptoms: A prospective study for potential diagnostic markers. *European Archives of Psychiatry and Clinical Neuroscience, 264*(7), 615–624. https://doi.org/10.1007/s00406-014-0494-8.

Kraus, T. A., Engel, S. M., Sperling, R. S., Kellerman, L., Lo, Y., Wallenstein, S., … Moran, T. M. (2012). Characterizing the pregnancy immune phenotype: Results of the viral immunity and pregnancy (VIP) study. *Journal of Clinical Immunology, 32*(2), 300–311. https://doi.org/10.1007/s10875-011-9627-2.

Kraus, T. A., Sperling, R. S., Engel, S. M., Lo, Y., Kellerman, L., Singh, T., … Moran, T. M. (2010). Peripheral blood cytokine profiling during pregnancy and postpartum periods. *American Journal of Reproductive Immunology, 64*(6), 411–426. https://doi.org/10.1111/j.1600-0897.2010.00889.x.

Le Doare, K., Holder, B., Bassett, A., & Pannaraj, P. S. (2018). Mother's milk: A purposeful contribution to the development of the infant microbiota and immunity. *Frontiers in Immunology, 9*, 361. https://doi.org/10.3389/fimmu.2018.00361.

Leff-Gelman, P., Mancilla-Herrera, I., Flores-Ramos, M., Cruz-Fuentes, C., Reyes-Grajeda, J. P., Garcia-Cuetara Mdel, P., … Pulido-Ascencio, D. E. (2016). The immune system and the role of inflammation in perinatal depression. *Neuroscience Bulletin, 32*(4), 398–420. https://doi.org/10.1007/s12264-016-0048-3.

Leonard, B., & Maes, M. (2012). Mechanistic explanations how cell-mediated immune activation, inflammation and oxidative and nitrosative stress pathways and their sequels and concomitants play a role in the pathophysiology of unipolar depression. *Neuroscience & Biobehavioral Reviews, 36*(2), 764–785. https://doi.org/10.1016/j.neubiorev.2011.12.005.

Li, J., Zhao, R., Li, X., Sun, W., Qu, M., Tang, Q., … Zhang, S. (2016). Shen-Qi-Jie-Yu-Fang exerts effects on a rat model of postpartum depression by regulating inflammatory cytokines and CD4(+)CD25(+) regulatory T cells. *Neuropsychiatric Disease and Treatment, 12*, 883–896. https://doi.org/10.2147/ndt.s98131.

Liu, H., Zhang, Y., Gao, Y., & Zhang, Z. (2016). Elevated levels of Hs-CRP and IL-6 after delivery are associated with depression during the 6 months post partum. *Psychiatry Research, 243*, 43–48. https://doi.org/10.1016/j.psychres.2016.02.022.

Maes, M., Lin, A. H., Ombelet, W., Stevens, K., Kenis, G., De Jongh, R., … Bosmans, E. (2000). Immune activation in the early puerperium is related to postpartum anxiety and depressive symptoms. *Psychoneuroendocrinology, 25*(2), 121–137.

Maes, M., Ombelet, W., De Jongh, R., Kenis, G., & Bosmans, E. (2001). The inflammatory response following delivery is amplified in women who previously suffered from major depression, suggesting that major depression is accompanied by a sensitization of the inflammatory response system. *Journal of Affective Disorders, 63*(1–3), 85–92.

Maes, M., Ombelet, W., Libbrecht, I. I., Stevens, K., Kenis, G., De Jongh, R., … Bosmans, E. (1999). Effects of pregnancy and delivery on serum concentrations of clara cell protein (CC16), an endogenous anticytokine: Lower serum CC16 is related to postpartum depression. *Psychiatry Research, 87*(2–3), 117–127.

Maestripieri, D., & Hoffman, C. L. (2011). Chronic stress, allostatic load, and aging in nonhuman primates. *Development and Psychopathology, 23*(4), 1187–1195. https://doi.org/10.1017/s0954579411000551.

Martin, S. M., Malkinson, T. J., Veale, W. L., & Pittman, Q. J. (1995). Fever in pregnant, parturient, and lactating rats. *American Journal of Physiology, 268*(4 Pt 2), R919–R923. https://doi.org/10.1152/ajpregu.1995.268.4.R919.

Miller, A. H., Maletic, V., & Raison, C. L. (2009). Inflammation and its discontents: The role of cytokines in the pathophysiology of major depression. *Biological Psychiatry, 65*(9), 732–741. https://doi.org/10.1016/j.biopsych.2008.11.029.

Mitchell, A. M., Porter, K., & Christian, L. M. (2018). Examination of the role of obesity in the association between childhood trauma and inflammation during pregnancy. *Health Psychology, 37*(2), 114–124. https://doi.org/10.1037/hea0000559.

Okun, M. L., Luther, J., Prather, A. A., Perel, J. M., Wisniewski, S., & Wisner, K. L. (2011). Changes in sleep quality, but not hormones predict time to postpartum depression recurrence. *Journal of Affective Disorders, 130*(3), 378–384. https://doi.org/10.1016/j.jad.2010.07.015.

Osborne, L. M., Brar, A., & Klein, S. L. (2018). The role of Th17 cells in the pathophysiology of pregnancy and perinatal mood and anxiety disorders. *Brain, Behavior, and Immunity*. https://doi.org/10.1016/j.bbi.2018.11.015.

Osborne, L. M., & Monk, C. (2013). Perinatal depression—the fourth inflammatory morbidity of pregnancy?: Theory and literature review. *Psychoneuroendocrinology, 38*(10), 1929–1952. https://doi.org/10.1016/j.psyneuen.2013.03.019.

Osborne, L. M., Yenokyan, G., Fei, K., Kraus, T., Moran, T., Monk, C., & Sperling, R. (2018). Innate immune activation and depressive and anxious symptoms across the peripartum: An exploratory study. *Psychoneuroendocrinology, 99*, 80–86. https://doi.org/10.1016/j.psyneuen.2018.08.038.

Payne, J. L., Palmer, J. T., & Joffe, H. (2009). A reproductive subtype of depression: Conceptualizing models and moving toward etiology. *Harvard Review of Psychiatry, 17*(2), 72–86. https://doi.org/10.1080/10673220902899706.

Pazos, M., Sperling, R. S., Moran, T. M., & Kraus, T. A. (2012). The influence of pregnancy on systemic immunity. *Immunologic Research, 54*(1–3), 254–261. https://doi.org/10.1007/s12026-012-8303-9.

Posillico, C. K., & Schwarz, J. M. (2016). An investigation into the effects of antenatal stressors on the postpartum neuroimmune profile and depressive-like behaviors. *Behavioural Brain Research, 298*(Pt B), 218–228. https://doi.org/10.1016/j.bbr.2015.11.011.

Saxena, V., Lienesch, D. W., Zhou, M., Bommireddy, R., Azhar, M., Doetschman, T., & Singh, R. R. (2008). Dual roles of immunoregulatory cytokine TGF-beta in the pathogenesis of autoimmunity-mediated organ damage. *The Journal of Immunology, 180*(3), 1903–1912.

Sherer, M. L., Posillico, C. K., & Schwarz, J. M. (2018). The psychoneuroimmunology of pregnancy. *Frontiers in Neuroendocrinology, 51*, 25–35. https://doi.org/10.1016/j.yfrne.2017.10.006.

Simpson, W., Steiner, M., Coote, M., & Frey, B. N. (2016). Relationship between inflammatory biomarkers and depressive symptoms during late pregnancy and the early postpartum period: A longitudinal study. *Revista Brasileira de Psiquiatria, 38*(3), 190–196. https://doi.org/10.1590/1516-4446-2015-1899.

Skalkidou, A., Sylvén, S. M., Papadopoulos, F. C., Olovsson, M., Larsson, A., & Sundström-Poromaa, I. (2009). Risk of postpartum depression in association with serum leptin and interleukin-6 levels at delivery: A nested case-control study within the UPPSAT cohort. *Psychoneuroendocrinology, 34*(9), 1329–1337. https://doi.org/10.1016/j.psyneuen.2009.04.003.

Smith, R. S. (1991). The macrophage theory of depression. *Medical Hypotheses, 35*(4), 298–306.

Voskuhl, R., & Momtazee, C. (2017). Pregnancy: Effect on multiple sclerosis, treatment considerations, and breastfeeding. *Neurotherapeutics, 14*(4), 974–984. https://doi.org/10.1007/s13311-017-0562-7.

Wesseloo, R., Kamperman, A. M., Bergink, V., & Pop, V. J. M. (2018). Thyroid peroxidase antibodies during early gestation and the subsequent risk of first-onset postpartum depression: A prospective cohort study. *Journal of Affective Disorders, 225*, 399–403. https://doi.org/10.1016/j.jad.2017.08.058.

CHAPTER 6

Phenomenology of perinatal obsessive–compulsive disorder

Jennifer L. Buchholz, Samantha N. Hellberg, Jonathan S. Abramowitz
Department of Psychology and Neuroscience, University of North Carolina at Chapel Hill, Chapel Hill, NC, United States of America

Introduction

For most expecting parents, pregnancy and the postpartum period are marked by excitement, wonder, anticipation, and jubilation. For some, however, this time is marked by the onset (or intensification) of severe emotional distress. Whereas a great deal of attention has been paid to symptoms of depression and psychosis that occur perinatally, much less consideration has been given to perinatal anxiety and to obsessive–compulsive disorder (OCD), which, though no longer considered an official anxiety disorder, is clearly closely related. Although early research suggested that pregnancy *protects* women from emotional disorders (Elliott, Rugg, Watson, & Brough, 1983), more recent evidence suggests that women are in fact at an *increased* risk of developing OCD during pregnancy and the postpartum period, with metaanalytic findings indicating rates of 2.07% and 2.43%, respectively (relative to 1.08% in the general population; Russell, Fawcett, & Mazmanian, 2013). Additionally, the majority of attention has focused on OCD symptoms in new and expecting *mothers* (perhaps due to prevailing biological explanations for OCD), despite research and clinical reports of OCD symptoms occurring with regularity among expecting and new *fathers* as well (Abramowitz, Moore, Carmin, Wiegartz, & Purdon, 2001; Abramowitz, Schwartz, & Moore, 2003). For the purposes of this book, however, we will focus on research on OCD in mothers.

The opening section of this chapter provides an overview and description of the clinical phenomenology of OCD, focusing on that occurring during pregnancy and the postpartum (collectively referred to as the perinatal period), including a case example to illustrate the cardinal features of this presentation or "subtype" of OCD. We then consider

Biomarkers of Postpartum Psychiatric Disorders
ISBN 978-0-12-815508-0
https://doi.org/10.1016/B978-0-12-815508-0.00006-0

comorbidity and differential diagnoses before turning to a review of the prevalence of perinatal OCD. A discussion of factors contributing to the maintenance of perinatal OCD is then presented, followed by an illustrative case example to conclude the chapter.

Clinical phenomenology

The nature of obsessive–compulsive disorder

OCD is described in the DSM-5 and the ICD-10 (American Psychiatric Association, 2013; World Health Organization, 1992) as a condition involving *obsessions* and/or *compulsions*. Obsessions are persistent intrusive thoughts, ideas, images, or doubts that are experienced as unacceptable, senseless, or bizarre. The intrusions also evoke subjective distress (e.g., anxiety, fear, doubt) and are not simply everyday worries about work, relationships, or finances. Common obsessions include exaggerated ideas of contamination, unwanted impulses to harm others, unfounded doubts about one's sexual preference, and intrusive sacrilegious images. Although highly individualistic, obsessions typically concern the following general themes: aggression and violence, responsibility for causing harm, contamination, sex, religion, and the need for exactness or completeness. Most individuals with OCD evidence multiple types of obsessions.

To control the anxiety caused by such thoughts, individuals with OCD attempt to avoid stimuli that trigger obsessions (e.g., public restrooms in the case of contamination obsessions). If such stimuli cannot be avoided, however, the person performs compulsive rituals—behavioral or mental acts that are completed according to self-generated "rules." The rituals are deliberate, yet clearly senseless or excessive in relation to the obsessional fear they are designed to neutralize (e.g., washing one's hands for 30 min after using the restroom). In this way, compulsive rituals in OCD differ from other sorts of repetitive behaviors that do not function as an escape from obsessions (e.g., repetitive hair pulling, compulsive gambling, tics). As with obsessions, rituals are highly individualized. Common *overt* rituals include excessive decontamination (e.g., washing), checking (e.g., locks, the stove), counting, and repeating routine actions (e.g., going through doorways). Examples of *covert* or *mental* rituals include excessive prayer and using special "safe" phrases or numbers to neutralize "unsafe" thoughts or stimuli (e.g., thinking the number 2 to "undo" the number 666). Obsessions and compulsions are functionally related in that obsessions (e.g., images of germs) *increase* subjective distress and rituals (e.g., washing) *reduce* distress.

Individuals with OCD display a range of insight into the senselessness of their symptoms, in that some acknowledge the irrationality of their obsessions and compulsions and others are firmly convinced that these symptoms are rational. Often, the degree of insight varies across time and obsessional themes. For example, one person might recognize obsessional thoughts of harm as senseless, yet have poor insight into the irrationality of contamination obsessions.

Perinatal obsessive–compulsive disorder symptoms

A number of studies have examined the presentation of OCD symptoms that begin or worsen specifically during pregnancy and the postpartum. Not surprisingly, in many cases, obsessions and compulsions appear strongly linked to pregnancy-related factors, such as responsibility for the care and protection of an otherwise helpless unborn or newborn child. Table 6.1 lists the characteristic signs and symptoms of this particular presentation of OCD.

Buttolph and Holland (1990) provided the first report of perinatal OCD. They described two women whose obsessions and compulsions began during pregnancy and were characterized by obsessional fear that the unborn baby would become ill from contamination by toxic agents. Both

Table 6.1 Characteristic signs and symptoms of perinatal obsessive-compulsive disorder.

Characteristic	Pregnancy	Postpartum
Onset	Gradual onset	Rapid onset
Content	Obsessional content involving contamination and illness	Obsessional content involving harm, accidents, or loss
Avoidance	Avoidance of contaminants	Avoidance of obsessional cues, sometimes including avoidance of the newborn
Rituals	Overt rituals (e.g., washing and checking)	Overt rituals (e.g., checking) Covert mental rituals, neutralizing
Comorbidity	Often associated with depressive symptoms	Often associated with depressive symptoms
Differentiation	Despite superficial similarity of thought content, not associated with postpartum psychosis	Despite superficial similarity of thought content, not associated with postpartum psychosis

women also had compulsive washing and cleaning rituals. These authors also described three cases in which rapid onset occurred soon *after* childbirth; postpartum symptoms concerned unwanted obsessional thoughts of harming the infant with a knife and of contaminating the baby. Sichel, Cohen, Dimmock, and Rosenbaum (1993) reported a case series of 15 women with postpartum OCD, all of whom reported intrusive, anxiety-evoking obsessional thoughts related to harming the infant (e.g., while the infant was sleeping). None of these patients evidenced overt compulsive rituals such as washing or checking; rather, they appeared to address their obsessions by avoiding the infant and triggers, such as potential weapons (e.g., knives).

In a second report, Sichel, Cohen, Rosenbaum, and Driscoll (1993) described two postpartum OCD cases that involved unwanted obsessional thoughts of stabbing the newborn. Both of these women also developed phobic avoidance of the child and of knives. Although neither reported observable rituals, covert (i.e., mental) rituals were not ruled out; as such, it is unclear whether or not these women performed other strategies (e.g., praying to themselves) in order to prevent feared outcomes and/or alleviate associated distress. Diaz et al., (1997) described five women with perinatal onset OCD. In three of these cases, where onset coincided with pregnancy, the major symptoms were contamination obsessions and washing or cleaning compulsions. For the remaining two women—who experienced onset soon *after* childbirth—the main symptoms were upsetting thoughts of harming (e.g., sexually violating or stabbing) the infant.

Thus, contamination obsessions and washing or cleaning rituals appear to predominate the OCD symptom picture among women who develop the condition *during* pregnancy; yet for those with onset *following* childbirth, unwelcome, distressing intrusive obsessional thoughts of harming the infant seem most common, though additional and larger studies are needed. Additionally, phobic avoidance of situations and stimuli that trigger such obsessions is frequent among affected individuals. However, there is little, if any, mention in the perinatal OCD literature of the perfectionism or symmetry/ordering symptoms that are often present in other populations of individuals with OCD.

Consistent with research on the occurrence of intrusive thoughts in the general population (e.g., Rachman & de Silva, 1978), postpartum obsessional intrusions may be either transitory or persistent and may even prompt actions designed to prevent feared consequences (Abramowitz et al., 2003; Miller, Hoxha, Wisner, & Gossett, 2015). For example, we treated a woman who spent several hours each day engaging in prayer rituals to neutralize obsessional thoughts that she would lose control and throw her infant in a dumpster. Another woman never allowed herself to be alone with her infant for fear that she might stab the child. Finally, clinical observations highlight the swift onset of postpartum obsessional symptoms, which is in contrast to the typically gradual onset of OCD symptoms that occurs during pregnancy or outside the context of pregnancy and childbirth altogether.

Comorbidity and differential diagnoses

Postpartum depression

There is a relationship between postpartum OCD symptoms and postpartum depression, particularly when the obsessions involve unwanted thoughts of hurting the newborn (Donahue Jennings, Ross, Popper, & Elmore, 1999; Miller et al., 2015; Wisner, Peindl, Gigliotti, & Hanusa, 1999). However, it is unknown whether the obsessions and compulsions serve as the cause or consequence of the depressive symptoms, or if symptoms develop concurrently. One study examining OCD symptoms among a large sample of treatment-seeking women with postpartum depression found that most participants endorsed at least some obsession-like thoughts and compulsion-like behaviors (Abramowitz et al., 2010). Moreover, there was an association between such OCD symptoms and depression severity. Given that depression often involves unwanted negative thoughts, it is possible that obsessional problems (e.g., unwanted aggressive thoughts) are symptoms of postpartum depression. Alternatively, it is plausible that the presence of distressing and recurrent intrusions gives rise to depressive symptoms (e.g., hopelessness, dysphoria, social isolation). Indeed, OCD in general is typically associated with secondary depressive symptoms (Nestadt et al., 2001).

About one-quarter to one-half of OCD sufferers in the general population also meet criteria for depression (Abramowitz, Storch, Keeley, & Cordell, 2007), and in the majority of cases, OCD symptoms predate the depression (Bellodi, Sciuto, Diaferia, Ronchi, & Smeraldi, 1992; Demal, Lenz, Mayrhofer, Zapotoczky, & Zitterl, 1993). Furthermore, a large proportion of women with postpartum OCD subsequently develop depression (Sichel, Cohen, Dimmock, 1993, Sichel, Cohen, Rosenbaum, 1993). In the context of postpartum depression, obsessions are common, with the content typically involving thoughts of harm related to the infant

(Donahue Jennings et al., 1999; ; Wenzel, Gorman, O'Hara, & Stuart, 2001; Wisner et al., 1999). To our knowledge, no formal diagnostic assessment of the development of OCD and depression from pregnancy to the postpartum period has been conducted. Such an assessment would provide much needed information about the timing of onset (i.e., OCD preceding depression, depression preceding OCD, vs. concurrent development) of these two conditions.

Postpartum psychosis

It is important to distinguish the symptoms of postpartum OCD from those of postpartum psychosis, considering that both conditions may involve ideation about harming the newborn. Despite the topographical similarity in thought *content*, there are crucial *functional* distinctions between the two conditions. That is, the ideation in psychosis is typically experienced as consistent with the person's delusional thinking and behavior (*ego-syntonic*). Accordingly, whereas obsessional thoughts in OCD evoke fear and compulsive rituals, delusional ideation in psychosis is not resisted or "neutralized" in this way. Rather, such ideation might be believed and even acted upon. Fortunately, hallucinations (e.g., "I saw bugs crawling all over the baby's skin") and delusions (e.g., "The baby is possessed by the devil"), which characterize psychosis, are quite rare in general, and usually absent in OCD. Moreover, hallucinations and delusions are usually accompanied by other "negative symptoms" of psychosis that are easily recognized such as loose associations, labile mood, and disorganization.

It is therefore important to keep in mind that obsessions in postpartum OCD (even those with violent or horrific content; e.g., the thought to put the baby in the microwave) are *not* associated with an increased risk of committing harm. This is because obsessional thoughts are experienced by the person as senseless, unwanted, and inconsistent with the person's sense of self (*ego-dystonic*). Individuals with OCD recognize that the intrusive thoughts are contrary to their better judgment and report a strong *fear* of engaging in unacceptable behavior (including fears of even *thinking about* engaging in such behavior). Moreover, individuals with postpartum OCD engage in disproportionate subjective resistance—avoidance and ritualizing in attempt to control or suppress obsessional thoughts—to reassure themselves that the feared consequences associated with their obsessions will not occur. In short, parents with postpartum OCD present with severe anxiety complaints (e.g., worry over whether or not they will harm), as opposed to psychotic symptoms.

Prevalence and incidence

Prenatal obsessive–compulsive disorder

A handful of studies have examined the incidence and prevalence of OCD in expecting parents using standard diagnostic interviews. Zar, Wijma, and Wijma (2002) assessed 506 women during the third trimester (week 32) of pregnancy using the Anxiety Disorders Interview Schedule—Revised. Fewer than 1% (0.2%) of the sample met criteria for OCD at the time of the interview. Sutter-Dallay Giaconne-Marcesche, Glatigny-Dallay, and Verdoux (2004) later interviewed 497 women using the Mini International Neuropsychiatric Interview for DSM-IV (Sheehan et al., 1998) in the third trimester (week 31) of pregnancy. They found a prenatal OCD prevalence of 1.2%. Uguz, Akman, Kaya, and Cilli (2007) collected data from 434 women in the third trimester of pregnancy using the Structured Clinical Interview for DSM-IV, Clinician Version (First, Williams, Karg, & Spitzer, 2015). The authors reported a prenatal OCD prevalence of 3.5% and an incidence of 0.5%. Most recently, a metaanalysis of 12 studies on the risk of OCD in pregnancy indicated a mean 2.07% prevalence rate (Russell et al., 2013).

Postpartum obsessive–compulsive disorder

Studies in which the incidence of postpartum OCD has been assessed provide evidence of an increased risk of OCD development and exacerbation during the postpartum period (Uguz, Akman et al., 2007; Uguz et al., 2007; Wenzel, Haugen, Jackson, & Brendle, 2005; Zambaldi et al., 2009). Wenzel et al. (2005) interviewed 147 women at 8 weeks postpartum using a semistructured diagnostic interview for *DSM-IV* (the Structured Clinical Interview for DSM or SCID), reporting a sample postpartum OCD incidence of 2.0%. Given that more than three-quarters (78%) of the sample were multiparous (i.e., had given birth to multiple children), and risk for postpartum OCD appears to be greatest for primiparous women (i.e., women who had only given birth to one child), this is likely an underestimate of the actual incidence of postpartum OCD among new mothers. For example, Uguz, Akman et al. (2007) conducted postpartum diagnostic assessments on 302 randomly sampled new mothers in Turkey at 6 weeks postpartum, also using the SCID. They reported an overall 4% ($n = 12$) incidence of postpartum OCD: 6.57% among primiparous women and 1.81% among multiparous women. Symptoms began by 4 weeks postpartum for all of the diagnosed women. Finally, Zambaldi et al. (2009)

assessed a sample of 400 women between 2 and 26 weeks postpartum using the Mini International Neuropsychiatric Interview (MINI) for *DSM-IV* and found a postpartum OCD incidence of 2.3% and a prevalence of 9%. Unfortunately, no information about the incidence among primiparous compared with multiparous women was provided. In their metaanalysis, Russell et al. (2013) found an overall prevalence rate of 2.43% across seven studies.

Maintaining factors

The etiology of OCD is largely unknown, and most likely involves a complex interaction of biological and learning/environmental factors. Nevertheless, research has revealed a number of factors that appear to play a role in the maintenance (if not the etiology) of OCD that occurs during the perinatal period (e.g., Fairbrother & Abramowitz, 2007). Indeed, the factors that maintain a problem are often distinct from those that lead to its onset in the first place. In this section, we discuss such phenomena.

Normally occurring intrusive thoughts

A large body of research indicates that unwanted intrusive thoughts—i.e., obsession-like ideas that intrude into consciousness, often provoking distress—are a normal and universal experience (e.g., Rachman & de Silva, 1978; Salkovskis & Harrison, 1984). Such thoughts often concern personally unacceptable topics that are frequently observed in OCD (e.g., harm, sex, blasphemy) and may be evoked by any manner of life events, current concerns, and present interests. Many people who have recently purchased a new automobile, for example, are prone to thinking about accidents. Current conceptual models of OCD view such intrusions as the seeds of clinical obsessions (e.g., Rachman, 1997).

Accordingly, researchers have examined unwanted intrusive thoughts as experienced by nonclinical samples of mothers (and fathers) during the perinatal and postpartum period (Abramowitz, Khandker, Nelson, Deacon, & Rygwall, 2006; Abramowitz et al., 2003; Fairbrother & Woody, 2008). Not surprisingly, these studies report that the vast majority of new and expecting parents also have such experiences. Fairbrother and Woody (2008), for example, found that 100% of a sample of 100 new mothers reported unwanted thoughts of accidental harm or death (e.g., suffocation) coming to the newborn, and nearly half of the sample reported thoughts of intentionally harming the child (e.g., verbally or physically abusing the child, leaving the baby somewhere, sexual molestation). Such prevalent intrusive thoughts are likely elicited by factors such as the stress of new parenthood and/or the added responsibility that accompanies this unique role (Fairbrother & Abramowitz, 2007). They may also occur as a function of psychological problems known to occur during the early postpartum period (e.g., depression and anxiety).

Situational and cognitive factors

Although the types of thoughts described above are normal and benign, certain aspects of the perinatal period in particular create especially fertile grounds for the development of obsessional problems by setting the stage for the misinterpretation of these thoughts as personally significant and threatening. This section discusses these factors.

Heightened responsibility. Pregnancy and childbirth evoke (especially for first-time parents) a sudden increase in feelings of responsibility for preventing harm. Circumstances that contribute to this include the abrupt increase in caregiving responsibilities and the simultaneous decrease in the role of healthcare providers immediately following childbirth. A number of authors have proposed that exaggerated estimates of responsibility for preventing harm are important in the maintenance of obsessional problems generally. Salkovskis (1985), for example, argued that the appraisal of normally occurring intrusive thoughts as indicating responsibility for causing or preventing harm leads to obsessional anxiety and compulsive behaviors characteristic of OCD. Thus, the perinatal period may uniquely expose first-time mothers (and fathers) to a chronic state of enhanced responsibility that elicits or aggravates OCD symptoms by facilitating misappraisals of intrusive thoughts.

Overestimates of threat. The tendency to overestimate the severity and likelihood of threat is also associated with OCD (e.g., Frost & Steketee, 2002). Indeed, pregnancy and early parenthood are characterized by an onslaught of information about the numerous potential risks for the pregnancy and the highly dependent, seemingly fragile, and vulnerable newborn (e.g., prenatally: miscarriage, contamination, premature labor, and birth; postpartum: car seats, sleeping habits, cribs, choking). Research demonstrates that individuals with OCD (including perinatal OCD) exaggerate such risks (e.g.,Abramowitz et al., 2006; Fairbrother, Thordarson, Challacombe, & Sakaluk, 2018; Frost & Steketee, 2002), and that such exaggerations longitudinally predict greater OCD symptoms in the postpartum period. Moreover, the perceived ubiquity of potentially threatening situations provides frequent triggers for intrusive thoughts of possible harm.

Misappraisals and overimportance of intrusive thoughts. The situational and cognitive factors described above can lead to new parents misinterpreting their normally occurring intrusive thoughts as highly significant and meaningful, especially if they have a tendency to believe that thoughts are morally equivalent to actions, or likely to lead to corresponding behavior (i.e., thought–action fusion; Shafran, Thordarson, & Rachman, 1996). Consider, for example, a new mother with an unwanted thought to shake her infant, who will not stop screaming. Whereas most parents would appraise this unsettling idea as meaningless (e.g., "I'm feeling angry, but I wouldn't do that"), a parent with postpartum OCD would misappraise this intrusion in a catastrophic way that evokes obsessional fear and doubt (e.g., "This thought means I am a dangerous person who must take extra precautions to ensure that I don't do anything terrible. What if I lose control and shake my baby!?"). To this end, in a longitudinal study, Abramowitz, Nelson et al. (2007) found that the tendency to interpret negatively the presence and meaning of unwanted intrusive infant-related thoughts at 1 month postpartum was predictive of greater OCD symptoms later in the postpartum, even after controlling for baseline levels of obsessions and compulsions.

Effects of avoidance and compulsive rituals

Misinterpretation of intrusive thoughts leads to obsessional fear and preoccupation, as well as a range of responses, such as avoidance, concealment of the thought, attempts to suppress the upsetting thought, and overt (e.g., checking) and covert (e.g., praying silently) compulsive rituals. All of these responses function to reduce obsessional distress as well as the perceived risk associated with the intrusive thought. For example, several times each night, when the intrusive thought of her infant dying of sudden infant death syndrome comes to mind, a new mother returns to the nursery to check that the baby is still alive. She puts her hand on the child and holds her own breath so that she can be sure that she feels the child breathing.

Such behavioral responses to upsetting obsessional thoughts are negatively reinforced by the immediate (albeit temporarily) reduction in distress they engender. However, they also increase the frequency of the unwanted obsessional thoughts by drawing undue attention toward the intrusions.

Moreover, the responses prevent the unambiguous disconfirmation of faulty appraisals of obsessional thoughts. That is, when no feared outcome occurs following avoidance or ritualistic behavior, the parent attributes this nonoccurrence directly to the behavior ("I didn't harm the baby because I said the prayer") instead of realizing that the intrusive thoughts are benign. Similarly, the individual who avoids or conceals his or her unwanted infant-related obsessions never learns that such thoughts are universal and safe experiences. This preserves the faulty beliefs pertaining to threat and responsibility and maintains the misinterpretation of otherwise harmless intrusive thoughts.

Numerous studies indicate that compulsive rituals play a role in the maintenance of OCD. An experimental study by Salkovskis, Thorpe, Wahl, Wroe, and Forrester (2003), for example, revealed that OCD patients instructed to perform covert compulsive rituals in response to an obsession reported more distress and stronger urges to ritualize, compared to patients instructed to count backwards from 20, when exposed to their most distressing obsessional thought.

Case vignette

"Diane," a 28 year-old public relations consultant, had given birth to her first child, "Ethan," 2 months previously. The pregnancy and delivery were routine and uncomplicated and Diane had no significant medical or psychiatric history. About 4 weeks after delivery, however, at one of Ethan's well-baby checkups, she informed her pediatrician that she was feeling "down" most of the day and sometimes felt afraid of interacting with Ethan. The pediatrician recognized these symptoms as postpartum depression and referred Diane to a clinical psychologist for evaluation. At her assessment, Diane described feeling depressed because she could not seem to stop "recurrent bad thoughts." Though she was embarrassed to disclose the nature of her thoughts at first, she later described thoughts of harming Ethan, including the idea of vigorously shaking him while her husband was at work. These thoughts were repugnant to Diane, who considered herself a "gentle person who had never harmed anyone." Indeed, she had no history of violent behavior and no intent to act on these intrusions. In fact, Diane spent undue amounts of time "fighting" these thoughts and trying to counteract them with more pleasant ideas. She also avoided being alone with her infant. Still, Diane said that trying to dismiss the ideas of hurting Ethan was not working, which caused her a great deal of distress. She reported thinking that because she could not get rid of these, she must be "going crazy" and therefore more likely to do something terrible. Diane's rituals and her avoidance of being alone with Ethan, for fear of "losing control" and harming him, were interfering with her family functioning as well as with her relationship with her newborn.

Summary and conclusions

As illustrated by this case vignette, OCD symptoms that develop or worsen perinatally are associated with significant distress and functional impairment. The content of the obsessions and nature of the compulsions are generally related to the protection of an unborn or newborn child, and most commonly involve contamination fears during pregnancy and intrusive thoughts about harm during the postpartum period. Perinatal OCD symptoms are maintained by the misinterpretation of intrusive thoughts as highly significant and meaningful, overestimates of threat and responsibility for preventing harm, compulsive rituals, and avoidance of fear cues.

Although research has documented an increased risk of OCD during pregnancy and the postpartum period, perinatal OCD is understudied and underrecognized relative to depression and psychosis. Additionally, the causal pathways through which perinatal OCD develops and cooccurs with perinatal depression remain largely unknown. Advances in empirical research on perinatal OCD, as well as an increased recognition of symptoms among clinicians, are essential for the development of more effective prevention and treatment programs. Furthermore, efforts to disseminate empirical knowledge regarding perinatal OCD to medical staff well-situated to educate patients on these concerns before they arise and to monitor, recognize, and refer patients to appropriate care during the postpartum period (e.g., obstetricians and gynecologists) may be particularly useful.

References

Abramowitz, J., Moore, K., Carmin, C., Wiegartz, P. S., & Purdon, C. (2001). Acute onset of obsessive-compulsive disorder in males following childbirth. *Psychosomatics, 42*(5), 429–431. https://doi.org/10.1176/appi.psy.42.5.429.

Abramowitz, J. S., Khandker, M., Nelson, C. A., Deacon, B. J., & Rygwall, R. (2006). The role of cognitive factors in the pathogenesis of obsessive-compulsive symptoms: A prospective study. *Behaviour Research and Therapy, 44*(9), 1361–1374. https://doi.org/10.1016/j.brat.2005.09.011.

Abramowitz, J. S., Meltzer-Brody, S., Leserman, J., Killenberg, S., Rinaldi, K., Mahaffey, B. L., & Pedersen, C. (2010). Obsessional thoughts and compulsive behaviors in a sample of women with postpartum mood symptoms. *Archives of Women's Mental Health, 13*(6), 523–530. https://doi.org/10.1007/s00737-010-0172-4.

Abramowitz, J. S., Nelson, C. A., Rygwall, R., & Khandker, M. (2007). The cognitive mediation of obsessive-compulsive symptoms: A longitudinal study. *Journal of Anxiety Disorders, 21*(1), 91–104. https://doi.org/10.1016/j.janxdis.2006.05.003.

Abramowitz, J. S., Schwartz, S. A., & Moore, K. M. (2003). Obsessional thoughts in postpartum females and their partners: Content, severity, and relationship with depression. *Journal of Clinical Psychology in Medical Settings, 10*(3), 157–164. https://doi.org/10.1023/A:1025454627242.

Abramowitz, J. S., Storch, E. A., Keeley, M., & Cordell, E. (2007). Obsessive-compulsive disorder with comorbid major depression: What is the role of cognitive factors? *Behaviour Research and Therapy, 45*(10), 2257–2267. https://doi.org/10.1016/j.brat.2007.04.003.

American Psychiatric Association. (2013). *Diagnostic and statistical manual of mental disorders* (5th ed.). Arlington, VA: American Psychiatric Publishing.

Bellodi, L., Sciuto, G., Diaferia, G., Ronchi, P., & Smeraldi, E. (1992). Psychiatric disorders in the families of patients with obsessive-compulsive disorder. *Psychiatry Research, 42*, 111–120.

Buttolph, M., & Holland, A. (1990). Obsessive-compulsive disorders in pregnancy and childbirth. In M. Jenike, L. Baer, & W. Minichiello (Eds.), *Obsessive compulsive disorders, theory and management* (2nd ed.). Chicago: Yearbook Medical Publishers.

Demal, U., Lenz, G., Mayrhofer, A., Zapotoczky, H.-G., & Zitterl, W. (1993). Obsessive-compulsive disorder and depression. *Psychopathology, 26*(3–4), 145–150. https://doi.org/10.1159/000284814.

Diaz, S. F., Grush, L. R., Sichel, D. A., & Cohen, L. S. (1997). Obsessive-compulsive disorder in pregnancy and the puerperium. In M. T. Pato, & G. Steketee (Eds.), *OCD across the life cycle* (pp. 97–112). Washington, DC: American Psychiatric Association Press.

Donahue Jennings, K., Ross, S., Popper, S., & Elmore, M. (1999). Thoughts of harming infants in depressed and nondepressed mothers. *Journal of Affective Disorders, 54*(1), 21–28. https://doi.org/10.1016/S0165-0327(98)00185-2.

Elliott, S. A., Rugg, A. J., Watson, J. P., & Brough, D. I. (1983). Mood changes during pregnancy and after the birth of a child. *British Journal of Clinical Psychology, 22*(4), 295–308. https://doi.org/10.1111/j.2044-8260.1983.tb00616.x.

Fairbrother, N., & Abramowitz, J. S. (2007). New parenthood as a risk factor for the development of obsessional problems. *Behaviour Research and Therapy, 45*(9), 2155–2163. https://doi.org/10.1016/j.brat.2006.09.019.

Fairbrother, N., Thordarson, D. S., Challacombe, F. L., & Sakaluk, J. K. (2018). Correlates and predictors of new mothers' responses to postpartum thoughts of accidental and intentional harm and obsessive compulsive symptoms. *Behavioural and Cognitive Psychotherapy, 46*(4), 437–453. https://doi.org/10.1017/S1352465817000765.

Fairbrother, N., & Woody, S. R. (2008). New mothers' thoughts of harm related to the newborn. *Archives of Women's Mental Health, 11*(3), 221–229. https://doi.org/10.1007/s00737-008-0016-7.

First, M., Williams, J., Karg, R., & Spitzer, R. (2015). *Structured clinical interview for DSM-5*. Arlington, VA: American Psychiatric Association.

Frost, R. O., & Steketee, G. S. (2002). *Cognitive approaches to obsessions and compulsions*. Amsterdam Netherlands: Elsevier.

Miller, E. S., Hoxha, D., Wisner, K. L., & Gossett, D. R. (2015). Obsessions and compulsions in postpartum women without obsessive compulsive disorder. *Journal of Women's Health, 24*(10), 825–830. https://doi.org/10.1089/jwh.2014.5063.

Nestadt, G., Samuels, J., Riddle, M. A., Liang, K. Y., Bienvenu, O. J., Hoehn-Saric, R., … Cullen, B. (2001). The relationship between obsessive-compulsive disorder and anxiety and affective disorders: Results from the Johns Hopkins OCD family study. *Psychological Medicine, 31*(3), 481–487.

Rachman, S. J. (1997). A cognitive theory of obsessions. *Behaviour Research and Therapy, 35*(9), 793–802.

Rachman, S. J., & de Silva, P. (1978). Abnormal and normal obsessions. *Behaviour Research and Therapy, 16*(4), 233–248.

Russell, E. J., Fawcett, J. M., & Mazmanian, D. (2013). Risk of obsessive-compulsive disorder in pregnant and postpartum women: A meta-analysis. *The Journal of Clinical Psychiatry, 74*(4), 377–385. https://doi.org/10.4088/JCP.12r07917.

Salkovskis, P. M. (1985). Obsessional-compulsive problems: A cognitive-behavioural analysis. *Behaviour Research and Therapy, 23*(5), 571–583. https://doi.org/10.1016/0005-7967(85)90105-6.

Salkovskis, P. M., & Harrison, J. (1984). Abnormal and normal obsessions: A replication. *Behaviour Research and Therapy, 22*(5), 549–552. https://doi.org/10.1016/0005-7967(84)90057-3.

Salkovskis, P. M., Thorpe, S. J., Wahl, K., Wroe, A. L., & Forrester, E. (2003). Neutralizing increases discomfort associated with obsessional thoughts: An experimental study with obsessional patients. *Journal of Abnormal Psychology, 112*(4), 709–715.

Shafran, R., Thordarson, D. S., & Rachman, S. (1996). Thought-action fusion in obsessive compulsive disorder. *Journal of Anxiety Disorders, 10*(5), 379–391. https://doi.org/10.1016/0887-6185(96)00018-7.

Sheehan, D. V., Lecrubier, Y., Sheehan, K. H., Amorim, P., Janavs, J., Weiller, E., … Dunbar, G. C. (1998). The mini-international neuropsychiatric interview (M.I.N.I): The development and validation of a structured diagnostic psychiatric interview for DSM-IV and ICD-10. *Journal of Clinical Psychiatry, 59*, 22–33.

Sichel, D. A., Cohen, L. S., Dimmock, J. A., & Rosenbaum, J. F. (1993). Postpartum obsessive compulsive disorder: A case series. *The Journal of Clinical Psychiatry, 54*(4), 156–159.

Sichel, D. A., Cohen, L. S., Rosenbaum, J. F., & Driscoll, J. (1993). Postpartum onset of obsessive-compulsive disorder. *Psychosomatics: Journal of Consultation and Liaison Psychiatry, 34*(3), 277–279. https://doi.org/10.1016/S0033-3182(93)71893-9.

Sutter-Dallay, A. L., Giaconne-Marcesche, V., Glatigny-Dallay, E., & Verdoux, H. (2004). Women with anxiety disorders during pregnancy are at increased risk of intense postnatal depressive symptoms: A prospective survey of the MATQUID cohort. *European Psychiatry, 19*(8), 459–463. https://doi.org/10.1016/j.eurpsy.2004.09.025.

Uguz, F., Akman, C., Kaya, N., & Cilli, A. S. (2007). Postpartum-onset obsessive-compulsive disorder: Incidence, clinical features, and related factors. *The Journal of Clinical Psychiatry, 68*(1), 132–138.

Uguz, F., Gezginc, K., Zeytinci, I. E., Karatayli, S., Askin, R., Guler, O., … Gecici, O. (2007). Obsessive-compulsive disorder in pregnant women during the third trimester of pregnancy. *Comprehensive Psychiatry, 48*(5), 441–445. https://doi.org/10.1016/j.comppsych.2007.05.001.

Wenzel, A., Gorman, L. L., O'Hara, M. W., & Stuart, S. (2001). The occurrence of panic and obsessive compulsive symptoms in women with postpartum dysphoria: A prospective study. *Archives of Women's Mental Health, 4*(1), 5–12. https://doi.org/10.1007/s007370170002.

Wenzel, A., Haugen, E. N., Jackson, L. C., & Brendle, J. R. (2005). Anxiety symptoms and disorders at eight weeks postpartum. *Journal of Anxiety Disorders, 19*(3), 295–311. https://doi.org/10.1016/j.janxdis.2004.04.001.

Wisner, K. L., Peindl, K. S., Gigliotti, T., & Hanusa, B. H. (1999). Obsessions and compulsions in women with postpartum depression. *The Journal of Clinical Psychiatry, 60*(3), 176–180.

World Health Organization. (1992). *The ICD-10 classification of mental and behavioural disorders: Clinical descriptions and diagnostic guidelines.* Geneva: World Health Organization.

Zambaldi, C. F., Cantilino, A., Montenegro, A. C., Paes, J. A., de Albuquerque, T. L. C., & Sougey, E. B. (2009). Postpartum obsessive-compulsive disorder: Prevalence and clinical characteristics. *Comprehensive Psychiatry, 50*(6), 503–509. https://doi.org/10.1016/j.comppsych.2008.11.014.

Zar, M., Wijma, K., & Wijma, B. (2002). Relations between anxiety disorders and fear of childbirth during late pregnancy. *Clinical Psychology & Psychotherapy, 9*(2), 122–130. https://doi.org/10.1002/cpp.305.

CHAPTER 7

Genetics of perinatal obsessive–compulsive disorder: a focus on past genetic studies to inform future inquiry

Lindsay R. Standeven, Gerald Nestadt, Jack Samuels
Department of Psychiatry and Behavioral Sciences at Johns Hopkins Medical Institutions, Baltimore, Maryland, United States of America

Introduction

While postpartum depression (PPD) and postpartum psychosis (PPP) have received a great deal of clinical, research, and media interest, other perinatal psychiatric illnesses—particularly perinatal obsessive–compulsive disorder (pOCD) or symptoms (pOCS)—have been relatively neglected. Accurate assessments of the prevalence of pOCD prevalence are limited, but recent estimates are that 2.07% of pregnant women and 2.48% of postpartum of women meet criteria for OCD and many more experience subthreshold obsessive and/or compulsive symptoms, in the peripartum period (Russell, Fawcett, & Mazmanian, 2013). As discussed elsewhere in this volume, there is mounting evidence that the onset of OCD may be linked to the reproductive cycle and preexisting hormonal vulnerability (Forray, Focseneanu, Pittman, McDougle, & Epperson, 2010), may carry significant morbidity risks, and is often underrecognized and inadequately treated (Brandes, Soares, & Cohen, 2004).

Despite the growing evidence that pOCD and pOCS carry serious risks for mother and baby, there are few studies characterizing the course of pOCD, nor assessing the comorbidity of pOCS and PPD. In contrast, the course and comorbidity of OCD and major depressive disorder (MDD) (i.e., not in the peripartum) is well established; individuals with OCD have an estimated lifetime comorbidity of MDD of greater than 70% (Murphy et al., 2013). There is growing evidence that the overlap between OCD and MDD is genetically driven. Prior studies have shown coaggregation of OCD and depression among family members of probands (Nestadt et al.,

Biomarkers of Postpartum Psychiatric Disorders
ISBN 978-0-12-815508-0
https://doi.org/10.1016/B978-0-12-815508-0.00007-2

2001), identified common serotonergic pathways, and provided estimates that more than half of the phenotypic overlap of OCD and MDD is genetically driven (Bolhuis et al., 2014). These findings support "genetic pleiotropy," in which specific genetic variants underlie more than one clinically defined psychiatric disorder.

To our knowledge, there are no known studies to date evaluating the heritability or genetic factors for pOCD nor pOCS. However, there is emerging evidence for the heritability of PPD, which, given the evidence for the genetic overlap of MDD and OCD, may help direct future research on the genetic basis of pOCD and pOCS. There is also considerable evidence demonstrating the genetic basis for OCD outside of the perinatal period, which may also help direct future research on pOCD or pOCS. Because the genetic basis of PPD is amply covered elsewhere in this volume (Chapter 2), we will focus in this chapter on an overview of the scientific literature supporting a genetic basis for OCD, examine possible overlaps between the literature on PPD and that on OCD, and offer suggestions for how to apply this existing research to the future research needed for pOCD.

Family and twin studies

The earliest findings demonstrating the genetic underpinnings of OCD emerged from family studies conducted in 1936. Lewis (1936) used the "family history method" to interview family members of 50 patients with OCD for mental illness and personality traits and found "pronounced obsessional traits" in 37% of parents and 21% of siblings. Building on these early data, researchers employed the "family study method," in which the prevalence of OCD and OCS in probands (i.e., index cases) and their relatives was compared to that in control families (non-OCD probands and family members). Pauls, Alsobrook, Goodman, Rasmussen, and Leckman (1995) found a higher risk for OCD (10%) and subthreshold OCD (8%) among relatives of probands than among controls (1.9% and 2.0%, respectively). Similarly, the Johns Hopkins Family Study found an increased lifetime prevalence of OCD among first-degree relatives of OCD cases compared to relatives of controls (12% vs. 2%) (Nestadt et al., 2000). Several other family studies have demonstrated markedly increased rates of OCD in relatives of adult and pediatric OCD cases (Hanna, Himle, Curtis, & Gillespie, 2005).

Twin studies provide insight into the relative contributions of genetic and environmental influences in OCD by comparing the concordance rates of OCD in monozygotic and dizygotic twins. Using data from the Maudsley Twin Registry, Carey and Gottesman (1981) found 87% concordance of "obsessive symptoms or features" in monozygotic twins as compared to 47% in dizygotic twins. Twin studies have also provided estimates of the "heritability" of OCD, i.e., the proportion of variation in dimensional obsessive–compulsive symptom scores that is explained by genetic effects, rather than by environmental influences. In two of the largest twin studies to date, Eley et al. (2003) and Hudziak et al. (2004) estimated the heritability of obsessive–compulsive behaviors to be 65% and 45%–61%, respectively. In a rigorous and comprehensive review of past twin studies, van Grootheest, Boomsma, Hettema, and Kendler (2008) estimated heritability to be 45%–65% among children and 27%–47% among adults.

Linkage studies

Once the heritability of an illness is observed, the next step is to identify specific candidate genes that contribute to the pathology. Genetic linkage analysis is employed to detect the chromosomal location of genes that colocalize (are linked) and therefore follow similar heritability patterns. By comparing the inheritance patterns of genetic loci between probands and controls, scientists are able to identify potential "genetic markers." Studying 56 relatives of pediatric probands across seven families, Hanna et al. (2002) were the first to identify a linkage peak with a maximum multipoint "logarithm of odds" (LOD) score of 2.24 in the 9p24 region of chromosome 9. This region on chromosome 9 is of great interest, having been previously associated with schizophrenia, bipolar affective disorder, and tic disorder (Abdulkadir et al., 2018; Afshari et al., 2015; Myles-Worsley et al., 2013). This finding has since been replicated by Willour et al. (2004). However, in the largest genome-wide linkage study to date, the 9p24 region was not significant (Shugart et al., 2006). Instead, Shugart et al. (2006) noted higher signals on chromosomes 1, 3, 6, 7, and 15, with the strongest evidence for linkage on chromosome 3, region 3q27-28 (LOD = 2.67). Yet other studies implicated increased genetic linkage associations among certain OCD subtypes (e.g., hoarding) (Samuels et al., 2007) or among subpopulations (males over females) (Wang et al., 2009). However, none have been replicated nor risen to the level of statistical significance, likely due to the low power of these studies for detecting what may be slight variations in a number of contributing genes.

Genome-wide association studies

Genome-wide association studies (GWAS) identify single nucleotide polymorphisms (SNPs) associated with certain symptoms or syndromes in large populations. The three largest GWAS of OCD to date identified candidate genes with promising functional implications of OCD. The Netherland National Twin Registry conducted a GWAS with a trend toward significance for a SNP in the myocyte enhancer factor 2B neighbor (MEF2BNB) on chromosome 19 (den Braber et al., 2016). The International Genetics Collaborative (IOCDF-GC) noted a trend within DLGAP1 (a scaffold protein of the postsynaptic density of glutamatergic neurons) and SNP rs6131295 near the BTBD3 gene (a regulator of dendritic orientation) (Stewart et al., 2013). Lastly, the OCD Collaborative Genetic Association Study (OCGAS) identified several candidate genes, none of which rose to the level of genome-wide significance. However, the smallest *P* value was associated with a SNP on chromosome 9 (as was previously noted in family linkage studies) near the protein tyrosine phosphate receptor gene (PTPRD) and is thought to be involved in the glutamatergic synapse regulation (Mattheisen et al., 2015). Altogether, GWAS have yet to establish statistical significance for any candidate genes for OCD, again likely due to limited power. Additionally, as with many other psychiatric illnesses, OCD (and pOCD by extension) is likely to follow a polygenic mode of inheritance, with multiple genes exerting a small effect and therefore requiring larger sample sizes to detect genetic variations related to illness versus those in the general population.

These studies may nevertheless point the way for future research in pOCD, as a number of the SNPs indicated in the early OCD studies have also been implicated in PPD. In the largest combined linkage and GWAS study to date in PPD, Mahon et al. (2009) studied subjects from the NIMH Genetics Initiative Bipolar Disorder project and the Genetics of Recurrent Early-Onset Depression study and found maximum linkage peaks for PPD symptoms on chromosome 1q21.3-q32.1 and 9p24.3-p22.3—thus identifying a region that is potentially implicated in both OCD and PPD and may serve as a link for future studies pOCD.

Candidate genes

SERT genes

Data from linkage and genome-wide association studies, as well as observations from pharmacologic clinical trials, have led to the identification of a number of candidate genes and theories of the pathophysiology of OCD. In particular, the combination of early studies implicating the role of serotonin in OCD and the observed clinical improvement of OCD symptoms using serotonin reuptake inhibitors catalyzed scientific inquiry to find candidate genes involved in the serotonergic system (Marazziti et al., 1999; Sinopoli, Burton, Kronenberg, & Arnold, 2017). Studies have implicated polymorphisms in the promoter region of the serotonin transporter gene (5-HTTLPR) and serotonin receptor subtype genes: the serotonin 2A receptor (5-HT2A), the serotonin 1B (5-HT1B or 5-HT1Dβ) receptor, and the serotonin 2C receptor (5-HT2C). In one of the most comprehensive metaanalyses, Taylor (2013) found that only polymorphisms in the 5HTR2A and 5-HTTLPR genes reached statistical significance for association with OCD. In particular, some studies have found that the long (L) allele (but not the short (S) allele) polymorphism of the serotonin transporter gene regulatory region is associated with OCD, as well as poor response to SSRIs and low blood serotonin levels among individuals with mood disorders generally (Bengel et al., 1999; Lesch et al., 1996; McDougle, Epperson, Price, & Gelernter, 1998). However, these findings have not been replicated in other studies and appear to vary across ethnicities (Kim & Kim, 2006).

Genetic polymorphisms in the SERT gene are well studied in the PPD genetic literature and may prove another bridge between the two disorders. In particular the 5-HTTLPR polymorphism has been associated with PPD onset within 6–8 weeks from delivery, but not thereafter (McEvoy, Osborne, Nanavati, & Payne, 2017).

Dopamine receptor and transporter genes

While alterations in serotonin transmission have been the predominant hypothesis for OCD pathogenesis, many patients do not respond to SSRI treatment. Instead, refractory cases sometimes benefit from augmentation with neuroleptic (antidopaminergic) agents (Agid & Lerer, 1999). Paradoxically, clozapine (a dopamine receptor 4 antagonist) can elicit obsessive–compulsive symptoms, making the role of dopaminergic pathways in the pathogenesis of OCD a key area for scientific inquiry. Unfortunately, the results of several studies investigating the association between OCD and alleles for the dopamine receptor 4 gene have been largely negative (Taylor, 2013). A minority have shown a positive association of dopamine receptor 4 only to specific OCD subtypes such as tic disorder and early-onset OCD. Other studies of dopamine receptor subtypes (DRD2 and DRD3) and the dopamine transporter (SLC6A3) have been negative. Still, the positive pharmacologic observations warrant continued scientific inquiry into dopamine receptor polymorphisms.

COMT gene

Investigations of the involvement of the catechol-O-methyltransferase (COMT) enzyme (involved in the degradation of adrenaline, noradrenaline, and dopamine) in OCD have produced intriguing results. A polymorphism in the COMT gene resulting in a substitution of valine (Val) for methionine (Met) leads to a reduction in enzymatic activity. Patients who are heterozygous for the alleles (Val/Met) have intermediate catabolism compared to homozygous individuals (Met/Met), who have the lowest COMT enzymatic activity (Harrison & Tunbridge, 2008). Karayiorgou et al. (1997) found that the lowest COMT activity was associated with a 2.58 relative risk for OCD compared to either the intermediate or highest functioning alleles among males (Karayiorgou et al., 1999). These data were replicated in a case control study by Pooley, Fineberg, and Harrison (2007), showing higher odds (odds ratio (OR) = 1.91, 95% confidence interval (CI) 1.07–3.40, $P = .026$) of the Met(158) allele in men with OCD but not women. A recent metaanalysis, pooling these data with those from other studies, estimated that the Met(158) allele conferred a 20% increased odds of OCD (OR = 1.20, CI: 1.001–1.438, $P = .010$) (Pooley et al., 2007; Taylor, 2013).

In addition, studies dating back to the 1970s have noted a correlation between COMT and estrogen regulation. In animal models, fluctuations in COMT are observed across the reproductive cycle, and in vitro studies have directly demonstrated the downregulation of COMT mRNA levels in a dose-dependent matter with treatment of 17B-estradiol in glial cells (Jiang, Xie, Ramsden, & Ho, 2003). The established role of COMT in depression, anxiety, and stress makes this an excellent candidate gene for perinatal mood disorders and pOCD. Indeed, recent genetic studies have noted that polymorphisms in the COMT gene yielding the low activity form (Met/Met) are positively associated with PPD occurring up to 6–8 weeks postpartum, but not thereafter (McEvoy et al., 2017).

Monoamine oxidase-A gene

The MAO-A (monoamine oxidase-A gene) locus is a prime candidate for the genetic susceptibility of OCD, given its role in COMT-related pathways, its colocalization with serotonergic neurons, and the fact that it provides at least 80% of MAO activity in the metabolic degradation of 5-hydroxytryptamine receptors (5-HT), norepinephrine (NE), dopamine (DE), and epinephrine (Karayiorgou et al., 1999). Building on prior research, Karayiorgou et al. (1999) observed that a genetic polymorphism resulting in a threonine (Thr) to glycine (Gly) substitution in exon 8 of MAO-A resulted in a higher enzymatic activity and was preferentially transmitted among male probands and particularly among males with comorbid OCD and MDD. An alternate polymorphism of the MAO-A gene, this time resulting in a silent mutation in exon 14 (EcoRV), also showed gender dimorphism; with substitution of a high-activity Thr being associated with OCD in males, whereas substitution of a low-activity cystine (Cys) allele was associated with OCD in females (Camarena et al., 2001; Lochner et al., 2004; Taylor, 2013).

The same high-activity polymorphism of MAO-A has been associated with MDD and therefore may play a role in PPD. Results from the three studies conducted in PPD to date have been mixed, with two studies showing a positive association between the high-activity allele and PPD at 6 weeks postpartum, but no association at other time points (McEvoy et al., 2017).

Glutamate transporter gene

As described earlier in this chapter, Hanna et al. (2002) found a linkage peak strongly associated with early-onset OCD on chromosome 9p24. This region includes the gene for the neuronal glutamate transporter (SCLA1) gene and plays a key role in glutamatergic transmission. SCLA1 is expressed throughout neurological areas (the cortex, striatum, and thalamus) that, when altered, may contribute to the pathogenesis of OCD. Multiple researchers to date have reproduced the association between SNPs in SCL1A1 (or flanking regions) and OCD (Dickel et al., 2006) and specifically among male probands (Arnold et al., 2017; Samuels et al., 2011). However, the metaanalysis completed by Taylor et al. (2013) did not support a significant association. To our knowledge, there are no current studies assessing genetic differences in the glutamate transporter among women with PPD or pOCD.

BDNF gene

There has also been interest in a potential role in OCD for brain-derived neurotrophic factor (BDNF). BDNF promotes neuronal proliferation, differentiation, and survival. BDNF has been proposed as a potential "biomarker" in mood disorders (namely, bipolar disorder and MDD), as low levels of serum BDNF have been observed in patients with MDD, which then subsequently increase following antidepressant treatment (Shimizu et al., 2003). Similar to COMT, substitution of Val for Met leads to a loss of function polymorphism (Val66Met) that has been associated with depression-like behaviors, increased stress, and impaired response to antidepressant treatments (e.g., fluoxetine and ketamine) in mouse models. Hall, Dhilla, Charalambous, Gogos, & Karayiorgou (2003) showed differing levels of BDNF haplotype transmission—with overtransmission of haplotype 2 and undertransmission of haplotype 4 (carrying the Met66 substitution)—in OCD probands. A similar pattern of undertransmission, or potentially, the loss of a "protective" effect of haplotype 4, was observed in prior studies (Hall et al., 2003). BDNF is also a promising gene candidate in the peripartum population. However, of the two studies completed to date among women with PPD, only one yielded a positive association among women with a psychiatric history and onset of symptoms prior to 6 weeks postpartum (McEvoy et al., 2017).

Oxytocin receptor

Oxytocin has a key role in the neuromodulatory effects of serotonin and dopamine as well as pathways relating to social development and reactivity. Early animal studies demonstrated a role for oxytocin in grooming behaviors in mice, which have since been used as a research analogue for compulsions (Van Wimersma Greidanus, Kroodsma, Pot, Stevens, & Maigret, 1990). In human studies, elevated CSF and plasma oxytocin levels correlate with OCD diagnosis and symptom severity (Leckman, 1994; Marazziti et al., 2015). There is also evidence suggesting that SSRI-induced reduction of obsessions may be in part mediated through the oxytocin pathway (Humble, Uvnas-Moberg, Engstrom, & Bejerot, 2013). In the most recent association study Kang, Kim, Kim, Hwang, and Kim (2017) examined 10 SNPs of the Oxytocin gene receptor (OXTR) and identified two significant SNPs of the OXTR that were predictive of later onset OCD, but not OCD status (vs. controls) nor symptom profiles.

Oxytocin also plays a key role in maternal behaviors and mother–infant attachment, which is often diminished in cases of PPD. Lower mid-pregnancy oxytocin levels have been associated with increased risk for PPD (Skrundz, Bolten, Nast, Hellhammer, & Meinlschmidt, 2011). The genetic literature examining differences in the oxytocin receptor polymorphism expression among women with PPD is limited and has yielded contradictory results (McEvoy et al., 2017). While the evidence is still sparse, the potential for oxytocin to play opposite roles in OCD versus PPD (promoting symptoms vs. protecting against them) makes it all the more important to study the role of oxytocin in pOCD.

Estrogen receptor genes

Given the growing evidence for sex differences in age of OCD onset (e.g., onset in males is at younger ages than in females), types of obsessions, comorbid illnesses (e.g., tic disorder), and associated genetic polymorphisms (e.g., MAO and COMT), researchers have hypothesized that estrogen receptor genes may play a role in genetic vulnerability to OCD though to date only one human genetic study has been undertaken. Alonso et al. (2011) examined differences in SNPs of ESR1 and ESR2 and found no significant difference between alleles in individuals with OCD versus controls. However, a number of SNPs on the ESR1 gene were associated with differences in obsessive and compulsive symptoms (e.g., higher rates of contamination obsessions). The role of genetic polymorphisms in genes encoding for gonadal steroids seems a logical place to begin studying the genetics of perinatal disorders, yet there are only two known genetic studies of ESR1 among women with PPD, only one of which showed a significant association with PPD (McEvoy et al., 2017), and ESR2 has not been studied at all.

Conclusion

Although research supporting the heritability for OCD started decades ago, there is still a tremendous amount of work needed to clarify how the proposed genetic candidates may play a role in the pathogenesis of OCD. Like many other psychiatric disorders, OCD appears to have a complex inheritance pattern in which many genes (rather than one genetic alteration) contribute to traits and subsequent risk for developing the disease syndrome. The most promising gene candidates for the pathogenesis of OCD are polymorphisms involving the glutamate transporter (SCL1A1), the serotonin transporter gene (5-HTTLPR), the serotonin receptor subtype 5HT2A, and COMT. Of these, the strongest candidate gene is the glutamate transporter (SLC1A1), as several research groups have reported associations between OCD and SNPs in this gene and found an association with SNPs about 9 kilobases "upstream" of the gene. However, the relevance of these findings to pOCD is unclear, as they were specific to families

Table 7.1 Summary of positive gene associations from postpartum depression and obsessive–compulsive disorder studies.

Genes	Evidence from PPD	Evidence from OCD
BDNF	Positive (in women with a psychiatric history at 6 weeks)	Negative
COMT Val58Met	Positive (among Met/Met polymorphism when PP defined as ≤8 weeks)	Positive (among males only)
Dopamine receptor subtypes: DR4	Negative	Negative
ESR	Positive (within 3 months)	Not assessed
Glutamate transporter—(SCLA1)	Not assessed	Negative
MAO-A	Positive (within 6 weeks PPD)	Negative
Oxytocin	Positive (8 weeks and 6 months)	Not assessed
SERT Gene: 5HTPLRP	Positive (majority ≤8 weeks PP, one study positive at 1 year)	Positive (among trialleic)
Serotonin Receptor Subtype: 5-HT2A	Not assessed	Positive

"Negative" indicates no positive findings as presented in Taylor et al. (2013) and McEvoy et al. (2017).
All the information in Table 7.1 was adapted from the findings provided in Taylor, S. (2013). Molecular genetics of obsessive-compulsive disorder: A comprehensive meta-analysis of genetic association studies. *Molecular Psychiatry*, 18(7), 799–805. https://doi.org/10.1038/mp.2012.76 and McEvoy, K., Osborne, L. M., Nanavati, J., & Payne, J. L. (2017). Reproductive affective disorders: A review of the genetic evidence for premenstrual dysphoric disorder and postpartum depression. *Current Psychiatry Reports*, 19(12), 94. https://doi.org/10.1007/s11920-017-0852-0.

with male probands, or to males affected with OCD (Sinopoli et al., 2017). Instead, further study of emerging evidence for sex differences in the genetics of OCD may be of greater utility in creating a model for understanding the pathogenesis of pOCD (Khramtsova et al., 2018).

The emerging literature on the genetics of PPD (see "General approach to genetic studies" section, Chapter 2) has shown similarly conflicting results. However, upon reviewing individual gene candidate studies, a notable pattern emerges in which there is a more consistent genetic basis for PPD when it is defined more narrowly. When the postpartum period is defined as occurring within 8 weeks of delivery, a number of gene candidates show robust and consistent results with the strongest evidence supporting a role of 5-HTTPLRP, MAO-A, COMT Val158Met polymorphisms. Findings of associations of PPD with BDNF, ESR-1, and OXTR have been inconsistent, perhaps due to limitations in study design and/or sample size. Of note, the strongest gene candidates are those previously implicated in MDD, suggesting that future studies on pOCD may want to similarly focus on gene candidates implicated in OCD more generally.

Further work will also be required to establish whether pOCD is a distinct subtype of OCD or not, and therefore whether the genetic etiology is distinct. Exciting areas of exploration remain untouched. Will we find that the same gene candidates implicated in OCD and PPD more generally are significant in studies among women with pOCD or pOCS? How will we define the timeline for pOCD to ensure that we are accurately studying genetically and hormonally "at-risk" women and not introduce additional cofounders? Although our understanding of the genetic underpinnings for pOCD has lagged behind the longstanding interest in the genetics of OCD, we can use the combination of improved scientific technology and research experience to work toward answering these and many other questions to elucidate the pathophysiology of pOCD and hopefully improve the treatment outcomes for the women and their children who suffer the consequence of these burdensome illnesses.

References

Abdulkadir, M., Londono, D., Gordon, D., Fernandez, T. V., Brown, L. W., Cheon, K. A., … Dietrich, A. (2018). Investigation of previously implicated genetic variants in chronic tic disorders: A transmission disequilibrium test approach. *European Archives of Psychiatry and Clinical Neuroscience, 268*(3), 301–316. https://doi.org/10.1007/s00406-017-0808-8.

Afshari, P., Myles-Worsley, M., Cohen, O. S., Tiobech, J., Faraone, S. V., Byerley, W., & Middleton, F. A. (2015). Characterization of a novel mutation in SLC1A1 associated with schizophrenia. *Molecular Neuropsychiatry, 1*(3), 125–144. https://doi.org/10.1159/000433599.

Agid, O., & Lerer, B. (1999). Risperidone augmentation of paroxetine in a case of severe, treatment-refractory obsessive-compulsive disorder without comorbid psychopathology. *Journal of Clinical Psychiatry, 60*(1), 55–56.

Alonso, P., Gratacos, M., Segalas, C., Escaramis, G., Real, E., Bayes, M., … Menchon, J. M. (2011). Variants in estrogen receptor alpha gene are associated with phenotypical expression of obsessive-compulsive disorder. *Psychoneuroendocrinology, 36*(4), 473–483. https://doi.org/10.1016/j.psyneuen.2010.07.022.

Arnold, P. D., Askland, K. D., Barlassina, C., Bellodi, L., Bienvenu, O. J., Black, D., … Zai, G. (2017). Revealing the complex genetic architecture of obsessive–compulsive disorder using meta-analysis. *Molecular Psychiatry, 23*(5), 1181–1188. https://doi.org/10.1038/mp.2017.154.

Bengel, D., Greenberg, B. D., Cora-Locatelli, G., Altemus, M., Heils, A., Li, Q., & Murphy, D. L. (1999). Association of the serotonin transporter promoter regulatory region polymorphism and obsessive-compulsive disorder. *Molecular Psychiatry, 4*(5), 463–466.

Bolhuis, K., McAdams, T. A., Monzani, B., Gregory, A. M., Mataix-Cols, D., Stringaris, A., & Eley, T. C. (2014). Aetiological overlap between obsessive-compulsive and depressive symptoms: A longitudinal twin study in adolescents and adults. *Psychological Medicine, 44*(7), 1439–1449. https://doi.org/10.1017/s0033291713001591.

Brandes, M., Soares, C. N., & Cohen, L. S. (2004). Postpartum onset obsessive-compulsive disorder: Diagnosis and management. *Archives of Women's Mental Health,* 7(2), 99–110. https://doi.org/10.1007/s00737-003-0035-3.

den Braber, A., Zilhao, N. R., Fedko, I. O., Hottenga, J. J., Pool, R., Smit, D. J., … Boomsma, D. I. (2016). Obsessive-compulsive symptoms in a large population-based twin-family sample are predicted by clinically based polygenic scores and by genome-wide SNPs. *Translational Psychiatry, 6*, e731. https://doi.org/10.1038/tp.2015.223.

Camarena, B., Rinetti, G., Cruz, C., Gomez, A., de La Fuente, J. R., & Nicolini, H. (2001). Additional evidence that genetic variation of MAO-A gene supports a gender subtype in obsessive-compulsive disorder. *American Journal of Medical Genetics, 105*(3), 279–282.

Carey, G., & Gottesman, II. (1981). Twin and family studies of anxiety, phobic, and obsessive disorders. In D. F. Klein, & J. G. Rabkin (Eds.), *Anxiety: New Research and Changing Concepts* (pp. 117–136). New York: Raven Press.

Dickel, D. E., Veenstra-VanderWeele, J., Cox, N. J., Wu, X., Fischer, D. J., Van Etten-Lee, M., … Hanna, G. L. (2006). Association testing of the positional and functional candidate gene SLC1A1/EAAC1 in early-onset obsessive-compulsive disorder. *Archives of General Psychiatry, 63*(7), 778–785. https://doi.org/10.1001/archpsyc.63.7.778.

Eley, T. C., Bolton, D., O'Connor, T. G., Perrin, S., Smith, P., & Plomin, R. (2003). A twin study of anxiety-related behaviours in pre-school children. *Journal of Child Psychology and Psychiatry, 44*(7), 945–960.

Forray, A., Focseneanu, M., Pittman, B., McDougle, C. J., & Epperson, C. N. (2010). Onset and exacerbation of obsessive-compulsive disorder in pregnancy and the postpartum period. *Journal of Clinical Psychiatry, 71*(8), 1061–1068. https://doi.org/10.4088/JCP.09m05381blu.

Hall, D., Dhilla, A., Charalambous, A., Gogos, J. A., & Karayiorgou, M. (2003). Sequence variants of the brain-derived neurotrophic factor (BDNF) gene are strongly associated with obsessive-compulsive disorder. *The American Journal of Human Genetics, 73*(2), 370–376. https://doi.org/10.1086/377003.

Hanna, G. L., Himle, J. A., Curtis, G. C., & Gillespie, B. W. (2005). A family study of obsessive-compulsive disorder with pediatric probands. *American Journal of Medical Genetics B Neuropsychiatric Genetics, 134b*(1), 13–19. https://doi.org/10.1002/ajmg.b.30138.

Hanna, G. L., Veenstra-VanderWeele, J., Cox, N. J., Boehnke, M., Himle, J. A., Curtis, G. C., … Cook, E. H., Jr. (2002). Genome-wide linkage analysis of families with obsessive-compulsive disorder ascertained through pediatric probands. *American Journal of Medical Genetics, 114*(5), 541–552. https://doi.org/10.1002/ajmg.10519.

Harrison, P. J., & Tunbridge, E. M. (2008). Catechol-O-methyltransferase (COMT): A gene contributing to sex differences in brain function, and to sexual dimorphism in the predisposition to psychiatric disorders. *Neuropsychopharmacology, 33*(13), 3037–3045. https://doi.org/10.1038/sj.npp.1301543.

Hudziak, J. J., Van Beijsterveldt, C. E., Althoff, R. R., Stanger, C., Rettew, D. C., Nelson, E. C., … Boomsma, D. I. (2004). Genetic and environmental contributions to the child behavior checklist obsessive-compulsive scale: A cross-cultural twin study. *Archives of General Psychiatry, 61*(6), 608–616. https://doi.org/10.1001/archpsyc.61.6.608.

Humble, M. B., Uvnas-Moberg, K., Engstrom, I., & Bejerot, S. (2013). Plasma oxytocin changes and anti-obsessive response during serotonin reuptake inhibitor treatment: A placebo controlled study. *BMC Psychiatry, 13*, 344. https://doi.org/10.1186/1471-244x-13-344.

Jiang, H., Xie, T., Ramsden, D. B., & Ho, S. L. (2003). Human catechol-O-methyltransferase down-regulation by estradiol. *Neuropharmacology, 45*(7), 1011–1018.

Kang, J. I., Kim, H. W., Kim, C. H., Hwang, E. H., & Kim, S. J. (2017). Oxytocin receptor gene polymorphisms exert a modulating effect on the onset age in patients with obsessive-compulsive disorder. *Psychoneuroendocrinology, 86*, 45–52. https://doi.org/10.1016/j.psyneuen.2017.09.011.

Karayiorgou, M., Altemus, M., Galke, B. L., Goldman, D., Murphy, D. L., Ott, J., & Gogos, J. A. (1997). Genotype determining low catechol-O-methyltransferase activity as a risk factor for obsessive-compulsive disorder. *Proceedings of the National Academy of Sciences of the USA, 94*, 4572–4575.

Karayiorgou, M., Sobin, C., Blundell, M. L., Galke, B. L., Malinova, L., Goldberg, P., … Gogos, J. A. (1999). Family-based association studies support a sexually dimorphic effect of COMT and MAOA on genetic susceptibility to obsessive-compulsive disorder. *Biological Psychiatry, 45*(9), 1178–1189.

Khramtsova, E. A., Heldman, R., Derks, E. M., Yu, D., Davis, L. K., & Stranger, B. E. (2018). Sex differences in the genetic architecture of obsessive-compulsive disorder. *American Journal of Medical Genetics B Neuropsychiatric Genetics*. https://doi.org/10.1002/ajmg.b.32687.

Kim, S. J., & Kim, C. H. (2006). The genetic studies of obsessive-compulsive disorder and its future directions. *Yonsei Medical Journal, 47*(4), 443–454. https://doi.org/10.3349/ymj.2006.47.4.443.

Leckman, J. F., Goodman, W. K., North, W. G., Chappell, P. B., Price, L. H., Pauls, D. L., … Cohen, D. J. (1994). Elevated cerebrospinal fluid levels of oxytocin in obsessive-compulsive disorder: Comparison with Tourette's syndrome and healthy controls. *Archives of General Psychiatry, 51*, 782–792.

Lesch, K. P., Bengel, D., Heils, A., Sabol, S. Z., Greenberg, B. D., Petri, S., … Murphy, D. L. (1996). Association of anxiety-related traits with a polymorphism in the serotonin transporter gene regulatory region. *Science, 274*(5292), 1527–1531.

Lewis, A. (1936). Problems of obsessional illness: (Section of psychiatry). *Proceedings of the Royal Society of Medicine, 29*(4), 325–336.

Lochner, C., Hemmings, S. M., Kinnear, C. J., Moolman-Smook, J. C., Corfield, V. A., Knowles, J. A., … Stein, D. J. (2004). Gender in obsessive-compulsive disorder: Clinical and genetic findings. *European Neuropsychopharmacology, 14*(2), 105–113. https://doi.org/10.1016/s0924-977x(03)00063-4.

Mahon, P. B., Payne, J. L., MacKinnon, D. F., Mondimore, F. M., Goes, F. S., Schweizer, B., … Potash, J. B. (2009). Genome-wide linkage and follow-up association study of postpartum mood symptoms. *American Journal of Psychiatry, 166*(11), 1229–1237. https://doi.org/10.1176/appi.ajp.2009.09030417.

Marazziti, D., Baroni, S., Giannaccini, G., Catena-Dell'Osso, M., Piccinni, A., Massimetti, G., & Dell'Osso, L. (2015). Plasma oxytocin levels in untreated adult obsessive-compulsive disorder patients. *Neuropsychobiology, 72*, 74–80.

Marazziti, D., Dell'Osso, L., Presta, S., Pfanner, C., Rossi, A., Masala, I., … Cassano, G. B. (1999). Platelet [3H]paroxetine binding in patients with OCD-related disorders. *Psychiatry Research, 89*(3), 223–228.

Mattheisen, M., Samuels, J. F., Wang, Y., Greenberg, B. D., Fyer, A. J., McCracken, J. T., … Nestadt, G. (2015). Genome-wide association study in obsessive-compulsive disorder: Results from the OCGAS. *Molecular Psychiatry, 20*(3), 337–344. https://doi.org/10.1038/mp.2014.43.

McDougle, C. J., Epperson, C. N., Price, L. H., & Gelernter, J. (1998). Evidence for linkage disequilibrium between serotonin transporter protein gene (SLC6A4) and obsessive compulsive disorder. *Molecular Psychiatry, 3*(3), 270–273.

McEvoy, K., Osborne, L. M., Nanavati, J., & Payne, J. L. (2017). Reproductive affective disorders: A review of the genetic evidence for premenstrual dysphoric disorder and postpartum depression. *Current Psychiatry Reports, 19*(12), 94. https://doi.org/10.1007/s11920-017-0852-0.

Murphy, D. L., Moya, P. R., Fox, M. A., Rubenstein, L. M., Wendland, J. R., & Timpano, K. R. (2013). Anxiety and affective disorder comorbidity related to serotonin and other neurotransmitter systems: Obsessive-compulsive disorder as an example of overlapping clinical and genetic heterogeneity. *Philosophical Transactions of the Royal Society of London B Biological Sciences, 368*(1615), 20120435. https://doi.org/10.1098/rstb.2012.0435.

Myles-Worsley, M., Tiobech, J., Browning, S. R., Korn, J., Goodman, S., Gentile, K., … Middleton, F. A. (2013). Deletion at the SLC1A1 glutamate transporter gene cosegregates with schizophrenia and bipolar schizoaffective disorder in a 5-generation family. *American Journal of Medical Genetics B Neuropsychiatric Genetics, 162b*(2), 87–95. https://doi.org/10.1002/ajmg.b.32125.

Nestadt, G., Samuels, J., Riddle, M., Bienvenu, O. J., 3rd, Liang, K. Y., LaBuda, M., … Hoehn-Saric, R. (2000). A family study of obsessive-compulsive disorder. *Archives of General Psychiatry, 57*(4), 358–363.

Nestadt, G., Samuels, J., Riddle, M. A., Liang, K. Y., Bienvenu, O. J., Hoehn-Saric, R., … Cullen, B. (2001). The relationship between obsessive-compulsive disorder and anxiety and affective disorders: Results from the Johns Hopkins OCD family study. *Psychological Medicine, 31*(3), 481–487.

Pauls, D. L., Alsobrook, J. P., 2nd, Goodman, W., Rasmussen, S., & Leckman, J. F. (1995). A family study of obsessive-compulsive disorder. *American Journal of Psychiatry, 152*(1), 76–84. https://doi.org/10.1176/ajp.152.1.76.

Pooley, E. C., Fineberg, N., & Harrison, P. J. (2007). The met(158) allele of catechol-O-methyltransferase (COMT) is associated with obsessive-compulsive disorder in men: Case-control study and meta-analysis. *Molecular Psychiatry, 12*(6), 556–561. https://doi.org/10.1038/sj.mp.4001951.

Russell, E. J., Fawcett, J. M., & Mazmanian, D. (2013). Risk of obsessive-compulsive disorder in pregnant and postpartum women: A meta-analysis. *Journal of Clinical Psychiatry, 74*(4), 377–385. https://doi.org/10.4088/JCP.12r07917.

Samuels, J. F., Bienvenu, O. J., 3rd, Pinto, A., Fyer, A. J., McCracken, J. T., Rauch, S. L., … Nestadt, G. (2007). Hoarding in obsessive-compulsive disorder: Results from the OCD collaborative genetics study. *Behaviour Research and Therapy, 45*(4), 673–686. https://doi.org/10.1016/j.brat.2006.05.008.

Samuels, J., Wang, Y., Riddle, M. A., Greenberg, B. D., Fyer, A. J., McCracken, J. T., … Nestadt, G. (2011). Comprehensive family-based association study of the glutamate transporter gene SLC1A1 in obsessive-compulsive disorder. *American Journal of Medical Genetics B Neuropsychiatric Genetics, 156b*(4), 472–477. https://doi.org/10.1002/ajmg.b.31184.

Shimizu, E., Hashimoto, K., Okamura, N., Koike, K., Komatsu, N., Kumakiri, C., … Iyo, M. (2003). Alterations of serum levels of brain-derived neurotrophic factor (BDNF) in depressed patients with or without antidepressants. *Biological Psychiatry, 54*(1), 70–75.

Shugart, Y. Y., Samuels, J., Willour, V. L., Grados, M. A., Greenberg, B. D., Knowles, J. A., … Nestadt, G. (2006). Genomewide linkage scan for obsessive-compulsive disorder: Evidence for susceptibility loci on chromosomes 3q, 7p, 1q, 15q, and 6q. *Molecular Psychiatry, 11*(8), 763–770. https://doi.org/10.1038/sj.mp.4001847.

Sinopoli, V. M., Burton, C. L., Kronenberg, S., & Arnold, P. D. (2017). A review of the role of serotonin system genes in obsessive-compulsive disorder. *Neuroscience and Biobehavioral Reviews, 80*, 372–381. https://doi.org/10.1016/j.neubiorev.2017.05.029.

Skrundz, M., Bolten, M., Nast, I., Hellhammer, D. H., & Meinlschmidt, G. (2011). Plasma oxytocin concentration during pregnancy is associated with development of postpartum depression. *Neuropsychopharmacology, 36*(9), 1886–1893. https://doi.org/10.1038/npp.2011.74.

Stewart, S. E., Yu, D., Scharf, J. M., Neale, B. M., Fagerness, J. A., Mathews, C. A., … Pauls, D. L. (2013). Genome-wide association study of obsessive-compulsive disorder. *Molecular Psychiatry, 18*(7), 788–798. https://doi.org/10.1038/mp.2012.85.

Taylor, S. (2013). Molecular genetics of obsessive-compulsive disorder: A comprehensive meta-analysis of genetic association studies. *Molecular Psychiatry, 18*(7), 799–805. https://doi.org/10.1038/mp.2012.76.

van Grootheest, D. S., Boomsma, D. I., Hettema, J. M., & Kendler, K. S. (2008). Heritability of obsessive-compulsive symptom dimensions. *American Journal of Medical Genetics B Neuropsychiatric Genetics, 147b*(4), 473–478. https://doi.org/10.1002/ajmg.b.30622.

Van Wimersma Greidanus, T. B., Kroodsma, J. M., Pot, M. L., Stevens, M., & Maigret, C. (1990). Neurohypophyseal hormones and excessive grooming behaviour. *European Journal of Pharmacology, 187*(1), 1–8.

Wang, Y., Samuels, J. F., Chang, Y. C., Grados, M. A., Greenberg, B. D., Knowles, J. A., … Nestadt, G. (2009). Gender differences in genetic linkage and association on 11p15 in obsessive-compulsive disorder families. *American Journal of Medical Genetics B Neuropsychiatric Genetics, 150b*(1), 33–40. https://doi.org/10.1002/ajmg.b.30760.

Willour, V. L., Yao Shugart, Y., Samuels, J., Grados, M., Cullen, B., Bienvenu, O. J., 3rd, … Nestadt, G. (2004). Replication study supports evidence for linkage to 9p24 in obsessive-compulsive disorder. *The American Journal of Human Genetics, 75*(3), 508–513. https://doi.org/10.1086/423899.

Further reading

Guglielmi, V., Vulink, N. C., Denys, D., Wang, Y., Samuels, J. F., & Nestadt, G. (2014). Obsessive-compulsive disorder and female reproductive cycle events: Results from the OCD and reproduction collaborative study. *Depression and Anxiety, 31*(12), 979–987. https://doi.org/10.1002/da.22234.

Guintivano, J., Arad, M., Gould, T. D., Payne, J. L., & Kaminsky, Z. A. (2013a). Antenatal prediction of postpartum depression with blood DNA methylation biomarkers. *Molecular Psychiatry, 19*, 633.

Kimmel, M., Clive, M., Gispen, F., Guintivano, J., Brown, T., Cox, O., … Kaminsky, Z. (2016). Oxytocin receptor DNA methylation in postpartum depression. *Psychoneuroendocrinology, 69*, 150–160. https://doi.org/10.1016/j.psyneuen.2016.04.008.

Kimmel, M., Kaminsky, Z., & Payne, J. L. (2013). Biomarker or pathophysiology? The role of DNA methylation in postpartum depression. *Epigenomics, 5*(5), 473–475. https://doi.org/10.2217/epi.13.51.

Osborne, L., Clive, M., Kimmel, M., Gispen, F., Guintivano, J., Brown, T., … Kaminsky, Z. (2016). Replication of epigenetic postpartum depression biomarkers and variation with hormone levels. *Neuropsychopharmacology, 41*(6), 1648–1658. https://doi.org/10.1038/npp.2015.333.

Osborne, L. M., Gispen, F., Sanyal, A., Yenokyan, G., Meilman, S., & Payne, J. L. (2017). Lower allopregnanolone during pregnancy predicts postpartum depression: An exploratory study. *Psychoneuroendocrinology, 79*, 116–121. https://doi.org/10.1016/j.psyneuen.2017.02.012.

CHAPTER 8

Hormonal contributions to perinatal obsessive–compulsive disorder

Mary Kimmel
Department of Psychiatry, University of North Carolina School of Medicine, Chapel Hill, NC, United States of America

Introduction

Obsessive–compulsive disorder (OCD) includes recurrent thoughts, urges, or mental images (obsessions) that are intrusive, unwanted, and cause distress and anxiety, along with repetitive behaviors (compulsions) designed to relieve the associated distress (Pauls, Abramovitch, Rauch, & Geller, 2014). OCD affects 0.6%–2.0% of adults (Crino, Slade, & Andrews, 2005; Speisman, Storch, & Abramowitz, 2011; Weissman et al., 1994). Symptoms of OCD are more common for women during periods of hormonal change. For example, a study of 57 women with OCD who completed a survey about their symptoms over time found that over 40% had premenstrual worsening and 13% of the women who had been pregnant had onset of OCD during pregnancy (Williams & Koran, 1997). In a survey of 101 female outpatients with OCD, almost half reported an exacerbation of symptoms during the premenstrual period and 17% reported an exacerbation during pregnancy. In contrast, 11% reported improvement of symptoms during pregnancy (Vulink, Denys, Bus, & Westenberg, 2006). The prevalence of OCD during the third trimester has been estimated at 3.5% of women (Uguz, Akman, Kaya, & Cilli, 2007), and two studies of the postpartum period found the incidence of postpartum OCD was 4% and the prevalence 9%, higher than in the general population (Uguz et al., 2007; Zambaldi et al., 2009). A recent metaanalysis, however, has proposed a lower prevalence for the perinatal period (though still elevated compared to the nonpregnant population) (Russell, Fawcett, & Mazmanian, 2013). Given the increase in prevalence of distressing obsessive–compulsive symptoms and OCD during periods of hormonal changes, including the perinatal period, hormonal sensitivity may play a role for some women in

Biomarkers of Postpartum Psychiatric Disorders
ISBN 978-0-12-815508-0
https://doi.org/10.1016/B978-0-12-815508-0.00008-4

developing either obsessive–compulsive symptoms as part of perinatal depression or frank perinatal OCD.

This targeted review will examine literature on the role of estrogen, progesterone, allopregnanolone, oxytocin, and cortisol in relation to perinatal obsessive–compulsive symptoms and OCD. A Pubmed search was completed for the terms "obsessive–compulsive disorder" and "hormones"; other articles as appropriate were also included in relation to topics that evolved from this search. The serotonin, dopamine, and glutamate systems have been implicated in the pathophysiology of OCD, and these systems will also be considered as they relate to estrogen and OCD.

Evidence for contribution by specific hormones

Estrogen

In a recent review of the literature, Karpinski et al. conclude that lack of estrogen, through modulation of neurotransmission signaling pathways, may lead to OCD symptoms (Karpinski, Mattina, & Steiner, 2017). They highlight a study in which administration of estradiol to prepubertal female rats decreased lever pressing (a compulsive-like behavior), while rapid estradiol withdrawal restored this behavior (Flaisher-Grinberg et al., 2009). Another group found that ovariectomized mice had an exacerbation of marble burying, another compulsive-like phenotype, as well as anxiety-like behaviors (as exhibited in the elevated plus maze and in open-field tests) when compared to mice with intact ovaries; administering estradiol attenuated the marble burying (Mitra, Bastos, Bates, Pereira, & Bult-Ito, 2016). Interestingly, there were no differences after ovariectomy in strains of mice with a tendency toward lower marble burying behaviors (Mitra et al., 2016). This indicates that both genetics and decreases in gonadal hormones are important to this compulsive behavior phenotype.

Other studies, however, have had contradictory findings. Low doses of 17 alpha-ethinyl estradiol, a common synthetic estrogen used in contraceptive pills, when given to pregnant mice, resulted in *greater* repetitive retrieval of the dam's own tail, possibly a compulsive-like behavior, beginning at the end of pregnancy and continuing during lactation (Catanese & Vandenberg, 2017). It is unclear, however, whether tail-retrieval is truly a compulsive-like behavior; it may be an adaptive behavior that prepares the dam for pup retrieval to the nest after the pups are delivered. These possibly contradictory findings may indicate that estrogen needs to be finely balanced to prevent behaviors that actually have some logical role in early motherhood from becoming compulsive-like. They may also highlight the heterogenous nature of OCD in the perinatal period, with estrogen playing very different roles in different types of OCD.

One mechanism by which estrogen changes lead to obsessive–compulsive symptoms and OCD is via estrogen's effects on serotonin synthesis, breakdown, and transmission, and on synapse assembly (Bethea & Reddy, 2012; Hernandez-Hernandez, Martinez-Mota, Herrera-Perez, & Jimenez-Rubio, 2018; Karpinski et al., 2017; Lokuge, Frey, Foster, Soares, & Steiner, 2011). Estradiol may increase serotonin synthesis by increasing expression of tryptophan hydroxylase (TPH), the rate-limiting enzyme in the production of serotonin, and specifically the form expressed in the central nervous system (Sanchez, Reddy, Centeno, Henderson, & Bethea, 2005). Estrogen has been found to increase TPH protein mass (Bethea, Mirkes, Shively, & Adams, 2000; Smith, Henderson, Abell, & Bethea, 2004). Therefore, the swift drop in estrogen at delivery could lead to less serotonin production through less active TPH, thus leading to a greater likelihood of obsessions and compulsions for vulnerable women. A drop in estrogen may also affect serotonin activity through increasing monoamine oxidase A (MAO-A), the enzyme that catalyzes serotonin into its inactive metabolite. Healthy women soon after birth have been found to have increased MAO-A density throughout the brain, as measured by PET scan (Sacher et al., 2010).

In addition to its effects on serotonin synthesis and breakdown, estrogen may also affect serotonin transport and signaling. Acute estradiol administration increased expression of the serotonin transporter, the protein involved in the reuptake of serotonin, in key brain regions such as the raphe nuclei, but chronic administration *decreased* serotonin transporter expression—meaning that in the pregnancy setting of chronic increased estrogen, women may have more serotonin available due to low reuptake (Karpinski et al., 2017; McQueen, Wilson, & Fink, 1997; Pecins-Thompson, Brown, & Bethea, 1998; Sumner et al., 1999). Estrogen may also downregulate the expression of autoreceptors that inhibit serotonin transmission, including serotonin-1A and serotonin-1B, and may increase transmission through serotonin-2A (Karpinski et al., 2017). In addition, agonists for the various serotonin transporters may behave differently in high and low estrogen settings. Agrati et al. found that a serotonin-1A agonist, 8-OH-DPAT, had differing effects on perseverative behavior based on timing in the menstrual cycle (Agrati et al., 2005). Middle-aged female mice in persistent diestrus with higher estradiol levels had higher perseverative behavior in a T-maze after being given the serotonin-1A agonist, 8-OH-DPAT,when compared to mice in diestrus but with irregular cycles and lower estradiol levels (Olvera-Hernández, Chavira, & Fernández-Guasti, 2013). The higher estradiol in persistent diestrus of middle-aged mice may indicate the limitations of comparing this model to humans in menopause, where estrogen levels are low. This work does, however, indicate that estrogen is very closely integrated with the serotonin system, and that alteration of serotonin transmission may affect perseverative behavior. Considerably more research is needed, however, to understand how this integration plays out in the different hormonal states of postpartum, perimenopause, and menopause.

Estrogen may also play a role in obsessive–compulsive symptoms through its modulation of the dopaminergic system. Dopamine agonists have been associated with worsening of obsessive–compulsive symptoms, and dopamine antagonists have been used for treatment of OCD (Eagle et al., 2014; Hirschtritt, Bloch, & Mathews, 2017; Koo, Kim, Roh, & Kim, 2010; Sesia, Bizup, & Grace, 2013). Estrogen may affect dopamine indirectly, by increasing the availability of serotonin, which in turn binds to receptors on dopamine neurons and inhibits them from releasing dopamine (Kapur & Remington, 1996). When a dopamine antagonist was given to younger mice, which were then followed up to adulthood, the long-term decrease in serotonin receptor expression in certain brain areas that was found in the males was attenuated in the females (De Santis, Huang, & Deng, 2018). This may indicate that estrogen contributes to the development of serotonin receptors even in the setting of early effects on the dopamine system. Estrogens also may affect dopamine levels without affecting serotonin by decreasing catechol-O-methyltransferase (COMT), an enzyme that metabolizes dopamine (Harrison & Tunbridge, 2008; Hill et al., 2007). These varied effects of estrogen on dopamine indicate that study of the dopamine system, and the prescription of dopamine antagonists for OCD, needs to consider sex and sex hormones as important variables in determining the impacts of these medications.

Another important neurotransmitter system implicated in OCD is the glutamatergic system (Wu, Hanna, Rosenberg, & Arnold, 2012). Higher levels of glutamate have been found in the cerebrospinal fluid (CSF) of OCD patients (Bhattacharyya et al., 2009; Chakrabarty, Bhattacharyya, Christopher, & Khanna, 2005). Estrogen regulates glutamate release as well as receptor and protein expression (Karpinski et al., 2017). Estrogen increases glutamate release, leading to the creation of new synapses (Schwarz,

Liang, Thompson, & McCarthy, 2008), but it may also at times downregulate glutamatergic release, ensuring that the system is not chronically overactivated (Karpinski et al., 2017; Kuo, Hariri, Bondar, Ogi, & Micevych, 2009). This role of up- and downregulation by estrogen is consistent with the complexity of many of estrogen's associations with OCD and may partially explain the differences in both timing and severity of obsessive–compulsive symptoms in the perinatal period.

Progesterone

It is difficult to disentangle the effects of progesterone from those of estrogen, with which it rises in tandem across pregnancy. One study found that progesterone given by itself reduced compulsive marble-burying in male mice (Umathe, Vaghasiya, Jain, & Dixit, 2009). There has been considerably less attention paid to progesterone's interactions with various neurotransmitter systems than there has been for estrogen, but some evidence indicates that it may also increase TPH and thus serotonin levels (Karpinski et al., 2017; Lu et al., 1999; Sanchez et al., 2005). Evidence about progesterone's effects on the dopamine system has been conflicting, and the evidence on progesterone's interactions with the glutamate system is sparse (Karpinski et al., 2017).

Allopregnanolone

Allopregnanolone is a metabolite of progesterone. Umathe et al. found that allopregnanolone decreased marble-burying behavior in male mice, with effects similar to those of fluoxetine (Umathe et al., 2009). In another model, 8-OH-DPAT, the serotonin-1A agonist, induced persistent compulsive-like behavior in rats, and acute and chronic administration of allopregnanolone decreased this persistent behavior (Umathe et al., 2009). Umathe et al. suggest that allopregnanolone may reduce dopamine release, increase firing of serotonergic neurons, and modulate the GABA-A receptor (Robichaud & Debonnel, 2006; Scheel-Krüger, 1986; Umathe et al., 2009). Dehydroepiandrosterone (DHEA), which has an opposite effect to allopregnanolone on GABA-A receptors, increased marble-burying (Compagnone & Mellon, 2000; Umathe et al., 2009). DHEA levels were higher in female OCD patients compared to controls (Erbay & Kartalci, 2015). Interestingly, brexanolone, an intravenous synthetic formulation of allopregnanolone, has been found to improve postpartum depression (Meltzer-Brody et al., 2018). As postpartum depression often has obsessive–compulsive symptoms, further research may find that interventions with allopregnanolone and GABA-A receptors may improve perinatal OCD.

Oxytocin

Evidence about oxytocin levels in OCD has been contradictory (Karpinski et al., 2017). In a study of lactating female mice compared to nonlactating and nulliparous females, lactating mice (with elevated oxytocin levels) buried fewer marbles and had lower compulsive-like nest building behavior; those given an oxytocin receptor antagonist had a reversal of this protection (Mitra, Mucha, Owen, & Bult-Ito, 2017). However, one study gave oxytocin as a treatment; in 12 subjects, including 9 women, intranasal oxytocin spray versus placebo did not reduce obsessive–compulsive symptoms (den Boer & Westenberg, 1992). Marazziti et al. show similar contradictory findings, with OCD subjects having higher oxytocin levels compared to controls, but lower oxytocin levels associated with greater symptom severity (Marazziti et al., 2015). Some of the contradictory findings from studies where oxytocin levels were increased could be explained if the increased oxytocin is in fact a compensatory mechanism.

In the Mitra et al. study, not only did lactating female rodents have less compulsive-like behavior at baseline— they also had the strongest response to the serotonin reuptake inhibitor (SSRI), fluoxetine, resulting in the lowest marble-burying behavior in lactating females compared to non-lactating and nulliparous females (Mitra et al., 2017). The serotonin and oxytocin systems are interconnected. Oxytocin levels correlate with serotonin transporter binding (Marazziti et al., 2012), so these two systems may work in concert to improve compulsions. For example, one study of 36 adults found that baseline oxytocin levels correlated with baseline severity of obsessive–compulsive symptoms, but only in those who would go on to respond to SSRIs (Humble, Uvnäs-Moberg, Engström, & Bejerot, 2013). Another study did not find differences in oxytocin following SSRI treatment in patients with major depressive disorder (MDD) (Keating, Dawood, Barton, Lambert, & Tilbrook, 2013). This may indicate that OCD is unique from MDD, that oxytocin is not as helpful in cases of OCD co-morbid with MDD, or that factors such as sex and age are important considerations (the study included males and females from ages 22–71 years). It is particularly important to clarify these relationships for perinatal OCD, given that breastfeeding, with its increased levels of oxytocin, may have an impact both on symptoms and on treatment.

The hypothalamic–pituitaryadrenal axis and cortisol

We have a fair amount of evidence for the involvement of the hypothalamic–pituitary–adrenal (HPA) axis with OCD. Subjects with OCD had greater HPA activity (Kluge et al., 2007) and higher urinary free cortisol when compared to controls, and both of these decreased with clomipramine treatment (Gehris, Kathol, Black, & Noyes, 1990). In another small study, cortisol levels were higher in 20 women with OCD than in 20 healthy controls (Erbay & Kartalci, 2015). Studying cortisol is complicated, as levels change across the day and in relation to stressors. A pilot study found that women with postpartum OCD had elevated baseline cortisol levels compared to controls and experienced more stress with a cold pressor test (Lord, Hall, Soares, & Steiner, 2011). A subsequent study employed a stressor in an MRI machine; while all women responded to the stressor, those with postpartum OCD had higher reported anxiety after the stressor, a trend toward a higher cortisol response, and a distinct pattern of brain activation in different regions when compared with postpartum women without OCD as well as healthy women outside the postpartum period (Lord, Steiner, Soares, Carew, & Hall, 2012). Another study found that reduced serotonin binding levels in the thalamus, as determined by PET, were associated with increased cortisol response and increased reported anxiety in nonperinatal subjects with MDD or OCD compared to controls (Reimold et al., 2011). In a study of 324 nonpostpartum participants, patients were categorized into those with MDD alone, those with OCD and MDD, those with OCD alone, and healthy controls. Participants' cortisol levels were studied upon awakening and across multiple time points across the day and in relation to dexamethasone suppression (Labad et al., 2018). Those with both OCD and MDD had a more flattened cortisol slope across the day, and OCD subjects with greater trait anxiety had increased cortisol levels upon awakening and reduced suppression with dexamethasone (Labad et al., 2018). Another two studies found that individuals of both sexes with both OCD with MDD had reduced dexamethasone suppression compared to healthy controls (Vallejo, 1998, Kuloglu, 2007), but a third study found reduced suppression only in men (Catapano, Monteleone, Maj, & Kemali, 1990). Labad et al. did not find differences in cortisol activity at different stages of the menstrual cycle. Cortisol rises throughout pregnancy and remains elevated into the postpartum period for a time before going back to prepregnancy levels. A recent review concluded that there are more studies finding no association between cortisol and antenatal depression than there are studies that do find an association (Orta, Gelaye, Bain, & Williams, 2018); no such studies have been conducted in perinatal OCD, and determining the role of the HPA axis will require research studies large enough to account for multiple factors and for heterogeneity in presentation. Perhaps the most important distinguishing factor will be whether or not perinatal women have comorbid depression.

Challenges

Both animal and human studies have implicated hormonal changes in obsessive–compulsive symptoms and OCD, but the exact mechanisms remain unclear. Association with estrogen is the most evident. However, interactions with serotonin, dopamine, and glutamate in concert with estrogen changes and OCD in the perinatal period are also important. The impact of progesterone is less clear. Studies of allopregnanolone are intriguing and a rich area for further investigation. Oxytocin has been implicated, but the evidence is contradictory, and it may be important chiefly because of its relation to the serotoninergic system. The HPA axis has been studied for many years in relation to OCD and there appears to be a relationship, but tests like the dexamethasone suppression test have not been consistent enough for use, and comorbidities and other factors likely complicate results.

One of the largest challenges in studying obsessive–compulsive symptoms and perinatal OCD is the heterogeneity of presentations and comorbidities. Those who develop new onset OCD during the perinatal period may differ from those who had preexisting OCD that worsens during the perinatal period. As noted, some women have worsening during pregnancy and others worsen postpartum; different systems, genetics, and other factors may lead to this heterogeneity. Postpartum depressive symptoms are highly comorbid with obsessive–compulsive symptoms, just as OCD is highly comorbid with MDD. Those with comorbid depressive symptoms may also differ from those with more predominant anxiety, those with autism spectrum symptoms, those with trichotillomania, or those with hoarding. There is a dopamine hypothesis for bipolar affective disorder, with a proposed mechanism of increased dopamine similar to a proposed mechanism in OCD (Ashok et al., 2017); the postpartum period is a period of special vulnerability for women with bipolar disorder, as it is for women with OCD. Symptoms such as obsessive cleaning, while traditionally associated with the contamination thoughts common in OCD, may also be associated with manic behavior. To date our diagnoses are based solely on groups of symptoms that in fact often cross over several diagnoses. Furthermore, each individual may have her own unique combination of factors (such as sensitivity to hormonal change, differences in multiple systems such as serotonin and dopamine signaling, amount of stressors or trauma, the functioning of the stress response, and genetic factors) that leads to her own particular vulnerability (Pauls et al., 2014).

Another challenge is the difficulty of studying hormonal changes. Hormones are constantly adapting and responding, and "normal" levels vary depending on circumstances, timing, environment, and other factors. Some studies have attempted to overcome this with controlled challenge tests that all participants undergo in a similar fashion. However, it is still difficult to account for all the individual factors. As a result, hormones have made poor biomarkers. Hormone levels may be more effective as part of a larger set of biomarkers that could better identify those at risk of perinatal OCD, give a broader picture of contributors to perinatal OCD, and indicate treatments and treatment response.

Future directions

Further study of larger numbers of subjects with postpartum depression including obsessive and compulsive symptoms and subjects with "pure" perinatal OCD may allow for studying obsessive and compulsive symptoms in relation to etiologies such as hormonal contributions. Larger studies are required to account for heterogeneity and to move toward a precision medicine approach to obsessive compulsive symptoms during the perinatal period.

Other novel areas of research may improve our understanding of hormones in relation to OCD and may serve as additional unique biomarkers. For example, Turna et al. have proposed that the microbiota-gut-brain axis is an important area of research in relation to OCD (Turna et al., 2016). In one study of a model inducing OCD-like behavior with a serotonin agonist, pretreatment with a probiotic prevented obsessive–compulsive behaviors including perseverative open-field locomotion, stereotypic turning, and marble burying, with results similar to pretreatment with an SSRI (Kantak, Bobrow, & Nyby, 2014). While this study used male mice only, the intriguing results introduce the question of whether probiotics affect the serotonin system and could be used to good effect in the perinatal period. Gut bacteria may also directly affect the development of perinatal OCD through hormonal manipulation. Certain gut bacteria secrete beta-glucuronidase, which deconjugates estrogen to its most bioactive form

(Baker, Al-Nakkash, & Herbst-Kralovetz, 2017). In the postpartum period, the presence of certain gut bacteria may allow for better balancing of the fluctuations in estrogen related to delivery (Rackers, Thomas, Williamson, Posey, & Kimmel, 2018). Similarly, during pregnancy, changes in gut microbiota may further exacerbate or decrease anxiety and be another important factor alongside unique responses to hormonal changes such as estrogen and cortisol. Gut bacteria have been found to associate with differences in the development and responsiveness of the HPA axis (Sudo et al., 2004). The gut bacteria could serve as a biomarker of hormonal systems and interactions and may also have the potential for treatment of OCD during periods of hormonal change such as the perinatal period. There is significant promise in better understanding the role of hormonal changes in relation to OCD and response to treatments for OCD. Novel areas of research such as inclusion of studies of the microbiota—gut—brain axis may further improve our ability to understand the role of hormones in OCD.

References

Agrati, D., Fernández-Guasti, A., Zuluaga, M. J., Uriarte, N., Pereira, M., & Ferreira, A. (2005). Compulsive-like behaviour according to the sex and the reproductive stage of female rats. *Behavioural Brain Research, 161*, 313—319. https://doi.org/10.1016/j.bbr.2005.02.017.

Ashok, A. H., Marques, T. R., Jauhar, S., Nour, M. M., Goodwin, G. M., Young, A. H., & Howes, O. D. (2017). The dopamine hypothesis of bipolar affective disorder: The state of the art and implications for treatment. *Molecular Psychiatry, 22*, 666—679. https://doi.org/10.1038/mp.2017.16.

Baker, J. M., Al-Nakkash, L., & Herbst-Kralovetz, M. M. (2017). Estrogen-gut microbiome axis: Physiological and clinical implications. *Maturitas, 103*, 45—53. https://doi.org/10.1016/j.maturitas.2017.06.025.

Bethea, C. L., Mirkes, S. J., Shively, C. A., & Adams, M. R. (2000). Steroid regulation of tryptophan hydroxylase protein in the dorsal raphe of macaques. *Biological Psychiatry, 47*, 562—576.

Bethea, C. L., & Reddy, A. P. (2012). Effect of ovarian steroids on gene expression related to synapse assembly in serotonin neurons of macaques. *Journal of Neuroscience Research, 90*, 1324—1334. https://doi.org/10.1002/jnr.23004.

Bhattacharyya, S., Khanna, S., Chakrabarty, K., Mahadevan, A., Christopher, R., & Shankar, S. K. (2009). Anti-brain autoantibodies and altered excitatory neurotransmitters in obsessive-compulsive disorder. *Neuropsychopharmacology: Official Publication of the American College Neuropsychopharmacology, 34*, 2489—2496. https://doi.org/10.1038/npp.2009.77.

den Boer, J. A., & Westenberg, H. G. (1992). Oxytocin in obsessive compulsive disorder. *Peptides, 13*, 1083—1085.

Catanese, M. C., & Vandenberg, L. N. (2017). Low doses of 17α-ethinyl estradiol alter the maternal brain and induce stereotypies in CD-1 mice exposed during pregnancy and lactation. *Reproductive Toxicology, 73*, 20—29. https://doi.org/10.1016/j.reprotox.2017.07.007.

Catapano, F., Monteleone, P., Maj, M., & Kemali, D. (1990). Dexamethasone suppression test in patients with primary obsessive-compulsive disorder and in healthy controls. *Neuropsychobiology, 23*, 53–56. https://doi.org/10.1159/000119427.

Chakrabarty, K., Bhattacharyya, S., Christopher, R., & Khanna, S. (2005). Glutamatergic dysfunction in OCD. *Neuropsychopharmacology: Official Publication of the American College Neuropsychopharmacology, 30*, 1735–1740. https://doi.org/10.1038/sj.npp.1300733.

Compagnone, N. A., & Mellon, S. H. (2000). Neurosteroids: Biosynthesis and function of these novel neuromodulators. *Frontiers in Neuroendocrinology, 21*, 1–56. https://doi.org/10.1006/frne.1999.0188.

Crino, R., Slade, T., & Andrews, G. (2005). The changing prevalence and severity of obsessive-compulsive disorder criteria from DSM-III to DSM-IV. *American Journal of Psychiatry, 162*, 876–882. https://doi.org/10.1176/appi.ajp.162.5.876.

De Santis, M., Huang, X.-F., & Deng, C. (2018). Early antipsychotic treatment in juvenile rats elicits long-term alterations to the adult serotonin receptors. *Neuropsychiatric Disease and Treatment, 14*, 1569–1583. https://doi.org/10.2147/NDT.S158545

Eagle, D. M., Noschang, C., d'Angelo, L.-S. C., Noble, C. A., Day, J. O., Dongelmans, M. L., … Robbins, T. W. (2014). The dopamine D2/D3 receptor agonist quinpirole increases checking-like behaviour in an operant observing response task with uncertain reinforcement: A novel possible model of OCD. *Behavioural Brain Research, 264*, 207–229. https://doi.org/10.1016/j.bbr.2013.12.040.

Erbay, L. G., & Kartalci, S. (2015). Neurosteroid levels in patients with obsessive-compulsive disorder. *Psychiatry Ivestigation, 12*, 538–544. https://doi.org/10.4306/pi.2015.12.4.538.

Flaisher-Grinberg, S., Albelda, N., Gitter, L., Weltman, K., Arad, M., & Joel, D. (2009). Ovarian hormones modulate "compulsive" lever-pressing in female rats. *Hormones and Behavior, 55*, 356–365. https://doi.org/10.1016/j.yhbeh.2008.10.002.

Gehris, T. L., Kathol, R. G., Black, D. W., & Noyes, R. (1990). Urinary free cortisol levels in obsessive-compulsive disorder. *Psychiatry Research, 32*, 151–158.

Harrison, P. J., & Tunbridge, E. M. (2008). Catechol-O-methyltransferase (COMT): A gene contributing to sex differences in brain function, and to sexual dimorphism in the predisposition to psychiatric disorders. *Neuropsychopharmacology: Official Publication of the American College Neuropsychopharmacology, 33*, 3037–3045. https://doi.org/10.1038/sj.npp.1301543.

Hernandez-Hernandez, O. T., Martinez-Mota, L., Herrera-Perez, J. J., & Jimenez-Rubio, G. (2018). Role of estradiol in the expression of genes involved in serotonin neurotransmission: Implications for female depression. *Current Neuropharmacology*, ISSN 1875-6190, PMID: 29956632. https://doi.org/10.2174/1570159X16666180628165107

Hill, R. A., McInnes, K. J., Gong, E. C. H., Jones, M. E. E., Simpson, E. R., & Boon, W. C. (2007). Estrogen deficient male mice develop compulsive behavior. *Biological Psychiatry, 61*, 359–366. https://doi.org/10.1016/j.biopsych.2006.01.012.

Hirschtritt, M. E., Bloch, M. H., & Mathews, C. A. (2017). Obsessive-compulsive disorder: Advances in diagnosis and treatment. *Journal of the American Medical Association, 317*, 1358–1367. https://doi.org/10.1001/jama.2017.2200.

Humble, M. B., Uvnäs-Moberg, K., Engström, I., & Bejerot, S. (2013). Plasma oxytocin changes and anti-obsessive response during serotonin reuptake inhibitor treatment: A placebo controlled study. *BMC Psychiatry, 13*, 344. https://doi.org/10.1186/1471-244X-13-344.

Kantak, P. A., Bobrow, D. N., & Nyby, J. G. (2014). Obsessive-compulsive like behaviors in house mice are attenuated by a probiotic (*Lactobacillus rhamnosus* GG). *Behavioural Pharmacology, 25*, 71–79. https://doi.org/10.1097/FBP.0000000000000013.

Kapur, S., & Remington, G. (1996). Serotonin-dopamine interaction and its relevance to schizophrenia. *American Journal of Psychiatry, 153*, 466–476. https://doi.org/10.1176/ajp.153.4.466.

Karpinski, M., Mattina, G. F., & Steiner, M. (2017). Effect of gonadal hormones on neurotransmitters implicated in the pathophysiology of obsessive-compulsive disorder: A critical review. *Neuroendocrinology, 105*, 1–16. https://doi.org/10.1159/000453664.

Keating, C., Dawood, T., Barton, D. A., Lambert, G. W., & Tilbrook, A. J. (2013). Effects of selective serotonin reuptake inhibitor treatment on plasma oxytocin and cortisol in major depressive disorder. *BMC Psychiatry, 13*, 124. https://doi.org/10.1186/1471-244X-13-124.

Kluge, M., Schüssler, P., Künzel, H. E., Dresler, M., Yassouridis, A., & Steiger, A. (2007). Increased nocturnal secretion of ACTH and cortisol in obsessive compulsive disorder. *Journal of Psychiatric Research, 41*, 928–933. https://doi.org/10.1016/j.jpsychires.2006.08.005.

Koo, M.-S., Kim, E.-J., Roh, D., & Kim, C.-H. (2010). Role of dopamine in the pathophysiology and treatment of obsessive-compulsive disorder. *Expert Review of Neurotherapeutics, 10*, 275–290. https://doi.org/10.1586/ern.09.148.

Kuloglu, M., Atmaca, M., Onal, S., Gecici, O., Bulut, V., & Tezcan, E. (2007). Neopterin levels and dexamethasone suppression test in obsessive-compulsive disorder. *Psychiatry Research, 151*(3), 265–270.

Kuo, J., Hariri, O. R., Bondar, G., Ogi, J., & Micevych, P. (2009). Membrane estrogen receptor-alpha interacts with metabotropic glutamate receptor type 1a to mobilize intracellular calcium in hypothalamic astrocytes. *Endocrinology, 150*, 1369–1376. https://doi.org/10.1210/en.2008-0994.

Labad, J., Soria, V., Salvat-Pujol, N., Segalàs, C., Real, E., Urretavizcaya, M., … Menchón, J. M. (2018). Hypothalamic-pituitary-adrenal axis activity in the comorbidity between obsessive-compulsive disorder and major depression. *Psychoneuroendocrinology, 93*, 20–28. https://doi.org/10.1016/j.psyneuen.2018.04.008.

Lokuge, S., Frey, B. N., Foster, J. A., Soares, C. N., & Steiner, M. (2011). Depression in women: Windows of vulnerability and new insights into the link between estrogen and serotonin. *Journal of Clinical Psychiatry, 72*, e1563–1569. https://doi.org/10.4088/JCP.11com07089.

Lord, C., Hall, G., Soares, C. N., & Steiner, M. (2011). Physiological stress response in postpartum women with obsessive-compulsive disorder: A pilot study. *Psychoneuroendocrinology, 36*, 133–138. https://doi.org/10.1016/j.psyneuen.2010.04.014.

Lord, C., Steiner, M., Soares, C. N., Carew, C. L., & Hall, G. B. (2012). Stress response in postpartum women with and without obsessive-compulsive symptoms: An fMRI study. *Journal of Psychiatry Neuroscience, 37*, 78–86. https://doi.org/10.1503/jpn.110005.

Lu, N. Z., Shlaes, T. A., Gundlah, C., Dziennis, S. E., Lyle, R. E., & Bethea, C. L. (1999). Ovarian steroid action on tryptophan hydroxylase protein and serotonin compared to localization of ovarian steroid receptors in midbrain of Guinea pigs. *Endocrine, 11*, 257–267. https://doi.org/10.1385/ENDO:11:3:257.

Marazziti, D., Baroni, S., Giannaccini, G., Betti, L., Massimetti, G., Carmassi, C., & Catena-Dell'Osso, M. (2012). A link between oxytocin and serotonin in humans: Supporting evidence from peripheral markers. *European Neuropsychopharmacology: The Journal of the European College of Neuropsychopharmacology, 22*, 578–583. https://doi.org/10.1016/j.euroneuro.2011.12.010.

Marazziti, D., Baroni, S., Giannaccini, G., Catena-Dell'Osso, M., Piccinni, A., Massimetti, G., & Dell'Osso, L. (2015). Plasma oxytocin levels in untreated adult obsessive-compulsive disorder patients. *Neuropsychobiology, 72*, 74–80. https://doi.org/10.1159/000438756.

McQueen, J. K., Wilson, H., & Fink, G. (1997). Estradiol-17 beta increases serotonin transporter (SERT) mRNA levels and the density of SERT-binding sites in female rat brain. *Brain Research. Molecular Brain Research, 45*, 13–23.

Meltzer-Brody, S., Colquhoun, H., Riesenberg, R., Epperson, C. N., Deligiannidis, K. M., Rubinow, D. R., … Kanes, S. (2018). Brexanolone injection in post-partum depression: Two multicentre, double-blind, randomised, placebo-controlled, phase 3 trials. *The Lancet, 392*, 1058–1070. https://doi.org/10.1016/S0140-6736(18)31551-4.

Mitra, S., Bastos, C. P., Bates, K., Pereira, G. S., & Bult-Ito, A. (2016). Ovarian sex hormones modulate compulsive, affective and cognitive functions in a non-induced mouse model of obsessive-compulsive disorder. *Frontiers in Behavioral Neuroscience, 10*, 215. https://doi.org/10.3389/fnbeh.2016.00215.

Mitra, S., Mucha, M., Owen, S., & Bult-Ito, A. (2017). Postpartum lactation-mediated behavioral outcomes and drug responses in a spontaneous mouse model of obsessive-compulsive disorder. *ACS Chemical Neuroscience, 8*, 2683–2697. https://doi.org/10.1021/acschemneuro.7b00231.

Olvera-Hernández, S., Chavira, R., & Fernández-Guasti, A. (2013). Sex- and endocrine-stage-differences in middle-aged rats in an animal model of OCD. *Progress in Neuro-Psychopharmacology and Biological Psychiatry, 44*, 81–87. https://doi.org/10.1016/j.pnpbp.2013.01.021.

Orta, O. R., Gelaye, B., Bain, P. A., & Williams, M. A. (2018). The association between maternal cortisol and depression during pregnancy, a systematic review. *Archives Women's Mental Health, 21*, 43–53. https://doi.org/10.1007/s00737-017-0777-y.

Pauls, D. L., Abramovitch, A., Rauch, S. L., & Geller, D. A. (2014). Obsessive-compulsive disorder: An integrative genetic and neurobiological perspective. *Nature Reviews Neuroscience, 15*, 410–424. https://doi.org/10.1038/nrn3746.

Pecins-Thompson, M., Brown, N. A., & Bethea, C. L. (1998). Regulation of serotonin re-uptake transporter mRNA expression by ovarian steroids in rhesus macaques. *Brain Research: Molecular Brain Research, 53*, 120–129.

Rackers, H. S., Thomas, S., Williamson, K., Posey, R., & Kimmel, M. C. (2018). Emerging literature in the microbiota-brain axis and perinatal mood and anxiety disorders. *Psychoneuroendocrinology, 95*, 86–96. https://doi.org/10.1016/j.psyneuen.2018.05.020.

Reimold, M., Knobel, A., Rapp, M. A., Batra, A., Wiedemann, K., Ströhle, A., … Heinz, A. (2011). Central serotonin transporter levels are associated with stress hormone response and anxiety. *Psychopharmacology, 213*, 563–572. https://doi.org/10.1007/s00213-010-1903-y.

Robichaud, M., & Debonnel, G. (2006). Allopregnanolone and ganaxolone increase the firing activity of dorsal raphe nucleus serotonergic neurons in female rats. *International Journal of Neuropsychopharmacology, 9*, 191–200. https://doi.org/10.1017/S146114570500595X.

Russell, E. J., Fawcett, J. M., & Mazmanian, D. (2013). Risk of obsessive-compulsive disorder in pregnant and postpartum women: A meta-analysis. *Journal of Clinical Psychiatry, 74*, 377–385. https://doi.org/10.4088/JCP.12r07917.

Sacher, J., Wilson, A. A., Houle, S., Rusjan, P., Hassan, S., Bloomfield, P. M., … Meyer, J. H. (2010). Elevated brain monoamine oxidase A binding in the early postpartum period. *Archives of General Psychiatry, 67*, 468–474. https://doi.org/10.1001/archgenpsychiatry.2010.32.

Sanchez, R. L., Reddy, A. P., Centeno, M. L., Henderson, J. A., & Bethea, C. L. (2005). A second tryptophan hydroxylase isoform, TPH-2 mRNA, is increased by ovarian steroids in the raphe region of macaques. *Brain Research. Molecular Brain Research, 135*, 194–203. https://doi.org/10.1016/j.molbrainres.2004.12.011.

Scheel-Krüger, J. (1986). Dopamine-gaba interactions: Evidence that GABA transmits, modulates and mediates dopaminergic functions in the basal ganglia and the limbic system. *Acta Neurologica Scandinavica – Supplement, 107*, 1–54.

Schwarz, J. M., Liang, S.-L., Thompson, S. M., & McCarthy, M. M. (2008). Estradiol induces hypothalamic dendritic spines by enhancing glutamate release: A mechanism for organizational sex differences. *Neuron, 58*, 584–598. https://doi.org/10.1016/j.neuron.2008.03.008.

Sesia, T., Bizup, B., & Grace, A. A. (2013). Evaluation of animal models of obsessive-compulsive disorder: Correlation with phasic dopamine neuron activity. *International Journal of Neuropsychopharmacology, 16*, 1295–1307. https://doi.org/10.1017/S146114571200154X.

Smith, L. J., Henderson, J. A., Abell, C. W., & Bethea, C. L. (2004). Effects of ovarian steroids and raloxifene on proteins that synthesize, transport, and degrade serotonin in the raphe region of macaques. *Neuropsychopharmacology: Official Publication of the American College Neuropsychopharmacology, 29*, 2035–2045. https://doi.org/10.1038/sj.npp.1300510.

Speisman, B. B., Storch, E. A., & Abramowitz, J. S. (2011). Postpartum obsessive-compulsive disorder. *Journal of Obstetric, Gynecologic and Neonatal Nursing, 40*, 680–690. https://doi.org/10.1111/j.1552-6909.2011.01294.x.

Sudo, N., Chida, Y., Aiba, Y., Sonoda, J., Oyama, N., Yu, X.-N., … Koga, Y. (2004). Postnatal microbial colonization programs the hypothalamic-pituitary-adrenal system for stress response in mice: Commensal microbiota and stress response. *Journal of Physiology, 558*, 263–275. https://doi.org/10.1113/jphysiol.2004.063388.

Sumner, B. E., Grant, K. E., Rosie, R., Hegele-Hartung, C., Fritzemeier, K. H., & Fink, G. (1999). Effects of tamoxifen on serotonin transporter and 5-hydroxytryptamine(2A) receptor binding sites and mRNA levels in the brain of ovariectomized rats with or without acute estradiol replacement. *Brain Research: Molecular Brain Research, 73*, 119–128.

Turna, J., Grosman Kaplan, K., Anglin, R., & Van Ameringen, M. (2016). What's bugging the gut in OCD?" A review of the gut microbiome in obsessive-compulsive disorder. *Depression and Anxiety, 33*, 171–178. https://doi.org/10.1002/da.22454.

Uguz, F., Akman, C., Kaya, N., & Cilli, A. S. (2007). Postpartum-onset obsessive-compulsive disorder: Incidence, clinical features, and related factors. *Journal of Clinical Psychiatry, 68*, 132–138.

Umathe, S. N., Vaghasiya, J. M., Jain, N. S., & Dixit, P. V. (2009). Neurosteroids modulate compulsive and persistent behavior in rodents: Implications for obsessive-compulsive disorder. *Progress in Neuro-Psychopharmacology and Biological Psychiatry, 33*(7), 1161–1166. https://doi.org/10.1016/j.pnpbp.2009.06.013.

Vallejo, J., Olivares, J., Marcos, T., Martinez-Osaba, M. J., Ribera, F., & Bulbena, A. (1988). Dexamethasone suppression test and primary obsessional compulsive disorder. *Comprehensive Psychiatry, 29*(5), 498–502.

Vulink, N. C. C., Denys, D., Bus, L., & Westenberg, H. G. M. (2006). Female hormones affect symptom severity in obsessive-compulsive disorder. *International Clinical Psychopharmacology, 21*, 171–175. https://doi.org/10.1097/01.yic.0000199454.62423.99.

Weissman, M. M., Bland, R. C., Canino, G. J., Greenwald, S., Hwu, H. G., Lee, C. K., … Wickramaratne, P. J. (1994). The cross national epidemiology of obsessive compulsive disorder. The Cross National Collaborative Group. *Journal of Clinical Psychiatry, 55*(Suppl. 1), 5–10.

Williams, K. E., & Koran, L. M. (1997). Obsessive-compulsive disorder in pregnancy, the puerperium, and the premenstruum. *Journal of Clinical Psychiatry, 58*, 330–334. quiz 335–336.

Wu, K., Hanna, G. L., Rosenberg, D. R., & Arnold, P. D. (2012). The role of glutamate signaling in the pathogenesis and treatment of obsessive-compulsive disorder. *Pharmacology Biochemistry and Behavior, 100*, 726–735. https://doi.org/10.1016/j.pbb.2011.10.007.

Zambaldi, C. F., Cantilino, A., Montenegro, A. C., Paes, J. A., de Albuquerque, T. L. C., & Sougey, E. B. (2009). Postpartum obsessive-compulsive disorder: Prevalence and clinical characteristics. *Comprehensive Psychiatry, 50*, 503–509. https://doi.org/10.1016/j.comppsych.2008.11.014

CHAPTER 9

Immunological biomarkers of postpartum obsessive–compulsive disorder

Lauren M. Osborne[1,2,3]
[1]Department of Psychiatry & Behavioral Sciences, The Johns Hopkins University School of Medicine, Baltimore, MD, United States of America; [2]Department of Gynecology & Obstetrics, The Johns Hopkins University School of Medicine, Baltimore, MD, United States of America; [3]Women's Mood Disorders Center, The Johns Hopkins University School of Medicine, Baltimore, MD, United States of America

Introduction

Obsessive-compulsive disorder (OCD) has long been suspected of having, at least in part, an immune etiology (da Rocha, Correa, & Teixeira, 2008; Dinn, Harris, McGonigal, & Raynard, 2001; Furtado & Katzman, 2015; Teixeira, Rodrigues, Marques, Miguel, & Fontenelle, 2014). The exact nature of this connection is as yet unclear, but recent systematic reviews support the idea that an immune-OCD connection may exist (Cosco et al., 2018; Marazziti, Mucci, & Fontenelle, 2018). One controversial hypothesis posits that antibodies directed toward streptococcal bacteria can cross-react with the basal ganglia, causing symptoms of OCD in children (pediatric autoimmune neuropsychiatric disorders associated with streptococcal infection, or PANDAS) (Dale, 2005). Some studies have found evidence of such antibodies and/or brain inflammation in children with OCD, while others have found none (Gause et al., 2009; Orlovska et al., 2017). Most studies have been small, with methodological limitations; a recent registry study found increased risk of OCD after any infection —thus supporting a role for immune activation, but not PANDAS specifically (Orlovska et al., 2017). Another hint toward the role of immune activation in OCD comes in its comorbidity with autoimmune disease (Sharma, Kline, Shreeve, & Hartman, 2011). One recent study found that individuals with OCD were 43% more likely to have any autoimmune illness than those without OCD, and first-degree relatives of people with OCD were also at increased risk of any autoimmune illness (Mataix-Cols et al., 2018).

These hints of immune involvement in general OCD make it reasonable to question whether immune system activation may account for the

Biomarkers of Postpartum Psychiatric Disorders
ISBN 978-0-12-815508-0
https://doi.org/10.1016/B978-0-12-815508-0.00009-6

elevated rates of obsessive–compulsive symptoms in the perinatal period (Russell, Fawcett, & Mazmanian, 2013), a time of known immune dysregulation. As outlined in greater detail in Chapter 5, we now know that there is considerable alteration in immune regulation across pregnancy and the postpartum, with the most recent research agreeing that innate immune barriers are strengthened in healthy pregnancies, but some aspects of adaptive immunity are less effective (Chen, Liu, & Sytwu, 2012; Holtan et al., 2015; Pazos, Sperling, Moran, & Kraus, 2012). Late in pregnancy, women are more vulnerable to pathogens, and autoimmune diseases also change during pregnancy (Leff-Gelman et al., 2016; Sherer, Posillico, & Schwarz, 2018), while the postpartum period sees a rebound of T-cell responses and autoimmune activity (Bergink et al., 2013; Wesseloo, Kamperman, Bergink, & Pop, 2018). If women are more vulnerable to OCD in the perinatal period, might these significant immune changes be in part responsible?

The purpose of this chapter is to review the evidence for immunological biomarkers of OCD in the perinatal period. (Although many of the chapters in this book are focused on the postpartum exclusively, I have chosen to expand to include pregnancy in this area, because the known data are so few.) In fact, there are no studies that point either toward or away from such evidence. There are, however, numerous studies outlining the role of immune activation in OCD outside the perinatal period—and a review of these, combined with a discussion about what we know about perinatal immune regulation, may point us toward future research directions. (Information on genetic pathways that may implicate immune-related genes can be found in Chapter 7 (Cappi et al., 2016, 2012; Hounie et al., 2008; Rodriguez, Morer, Gonzalez-Navarro, Gasso, et al., 2017).)

Autoantibodies and OCD

A logical place to start in asking about the immune connection to OCD is to ask whether OCD might be an autoimmune illness. Autoantibodies directed at the basal ganglia and thalamus have been found to be significantly higher in the cerebral spinal fluid (but not the serum) of OCD subjects compared to healthy controls, and both glutamate and glycine levels were also higher in OCD subjects, perhaps implicating autoantibodies in affecting the levels of neurotransmitters (Bhattacharyya et al., 2009). Antibasal ganglia antibodies were also found in a much higher proportion of OCD subjects when compared to healthy controls (Nicholson et al.,

2012). A recent metaanalysis pooled seven case-control studies involving 844 participants and found a consistently higher proportion of antibasal ganglia antibodies in subjects with OCD when compared to healthy controls (Pearlman, Vora, Marquis, Najjar, & Dudley, 2014). Another study examined antibrain antibodies in subjects with OCD compared to those with major depressive disorder (MDD) and found that a higher proportion of the OCD subjects had antibodies (Maina et al., 2009). Serum antibodies against dopamine receptors and lysoganglioside were elevated in youth with OCD compared to healthy subjects (Cox et al., 2015). Some studies have looked to specific infections for the development of autoantibodies, finding a potential association with *Toxoplasma gondii* IgG antibodies in those with OCD (Miman et al., 2010). Overall, then, elevated autoantibodies have been implicated in patients with OCD—but all of these studies are small, and none has been carried out in the perinatal period.

Inflammatory biomarkers and OCD

The earliest work in psychoneuroimmunology concerned levels of mostly peripheral cytokines (though more recent work has called into question the utility of such an approach), and OCD has been no exception. Various studies have examined individual inflammatory biomarkers, in either peripheral blood or cerebral spinal fluid, in subjects with OCD. One of the earliest studies found no difference in plasma levels of cytokines including interleukin (IL)-1β, IL-6, soluble IL-6 receptor (sIL-6R), sIL-2R, and transferrin receptor (TfR), but did find that IL-6 or sIL-6R concentrations were correlated to the severity of compulsive (but not obsessive) symptoms (Maes, Meltzer, & Bosmans, 1994). Another early study found no difference in IL-1β or IL-6, but did find lower plasma levels of TNF-α in OCD subjects compared to healthy controls (Monteleone, Catapano, Fabrazzo, Tortorella, & Maj, 1998). No differences were found in CSF IL-6 between OCD subjects and age- and sex-matched healthy controls in another small study (Carpenter et al., 2002). On the other hand, OCD subjects had increased levels of plasma CCL3, CXCL8, sTNFR1, and sTNFR2 when compared to healthy controls in another small study (Fontenelle et al., 2012) and increased plasma levels of IL-6 and TNF-α in study that found modest correlations between these two cytokines and duration of illness and age of onset (negative), respectively (Konuk et al., 2007). Increases in IL-2, IL-4, IL-6, IL-10, and TNF-α levels have also been found in OCD subjects compared to healthy controls (Rao et al., 2015). There have been a number of metaanalyses and systematic reviews as well, including one that found consistently decreased levels of IL-1β, as well as differences in levels of IL-6 and TNF-α that were affected by age and medication use (Gray & Bloch, 2012), in OCD subjects compared to healthy controls. Another recent systematic review and metaanalysis concluded overall that the immune association with OCD remains unproven—pooled results were negative for associations with IL-6, TNF-α, IL-1β, IL-4, and IL-10, but methodological differences among studies (including in key points such as medication use) make it impossible at this time to rule out an immune connection (Cosco et al., 2018).

The above studies examined levels of circulating cytokines, with findings that are contradictory and in sum give us no clear picture of whether cytokines are altered in OCD. Other studies have looked instead at whether immune cells of OCD subjects, when stimulated with an immune threat, produce differing amounts of cytokines than do cells of non-OCD subjects. This may be a more reliable method of determining difference, as circulating levels can be affected by many different elements. One such study found that the cells of OCD subjects, when stimulated with lipopolysaccharide (LPS), produced lower amounts of IL-6 than did the cells of both healthy controls and those with social anxiety disorder (S. Fluitman et al., 2010). This same group also looked at stimulated cytokines when OCD subjects and healthy controls were exposed to disgust-related objects and found that the OCD subjects produced lower levels of both IL-6 and TNF-α than did healthy controls (S. B Fluitman, Denys, Heijnen, & Westenberg, 2010). These results, while sparse, may indicate that there is a difference in the immune response to stimuli in patients with OCD.

Leukocyte subpopulations and OCD

Few studies have gone beyond peripheral markers, stimulated or not, to look at leukocyte subpopulations—a method that may yield information that is less susceptible to noise than even stimulated cytokines. Neutrophils have been found to be decreased in individuals with OCD compared to healthy controls in one small study (Atmaca, Kilic, Koseoglu, & Ustundag, 2011); others have found that $CD8^+$ lymphocytes are increased, and $CD4^+$ lymphocytes decreased, in OCD subjects compared with healthy controls (Marazziti et al., 1999). Natural killer cells were found to be elevated in male OCD subjects and those with MDD when compared to healthy controls; the elevation normalized after treatment for those with MDD but not those with OCD (Ravindran, Griffiths, Merali, & Anisman, 1999). A pediatric population study showed higher proportions of monocytes (and, when stimulated, of the cytokines they produce) in OCD subjects compared to healthy controls (Rodriguez, Morer, Gonzalez-Navarro, Serra-Pages, et al., 2017). On the other hand, no differences were found when T cell subsets were evaluated in untreated OCD patients compared to healthy controls and to the same patients after treatment with clomipramine (Barber et al., 1996). More recent research has highlighted a possible role for T_H17 cells. A mouse model of autoimmune encephalitis displayed OCD-like behaviors, which improved upon depletion of T_H17 cells; the same behavior could be induced by adoptive transfer of T_H17 cells in healthy mice (Kant, Pasi, & Surolia, 2018). In sum, a number of small studies have found differences in leukocyte subpopulations when comparing subjects with OCD to either healthy controls or subjects with other psychiatric illnesses—but with so few studies, focusing on such different aspects of immune cells, we have no clear answers.

Treatment effects on immune parameters

Although the above research makes it clear that we do not yet know how and in what way the immune system is related to OCD, a few studies have nevertheless examined whether standard treatments for OCD may be affecting symptomatology through effects on the immune system. No difference in leukocyte subpopulations was found when comparing untreated OCD subjects to the same subjects after treatment with clomipramine (Barber et al., 1996; Weizman et al., 1996). Treatment with paroxetine and venlafaxine did not affect the ex vivo production of multiple immune parameters (including both inflammatory cytokines and leukocyte subpopulations) in a small group of subjects with OCD, though those who were considered responders to the medication (with at least a 25% decrease in scores on the Y-BOCS) differed from nonresponders in their change in TNF-α (Denys, Fluitman, Kavelaars, Heijnen, & Westenberg, 2006). The group (cited above) that found an increase of $CD8^+$ lymphocytes and a decrease in $CD4^+$ lymphocytes at baseline in OCD subjects when compared to healthy controls found that these proportions normalized after treatment with SRIs (Marazziti et al., 2009). This amount of evidence is too little to say whether or not treatment can affect immune parameters in OCD subjects.

OCD in the perinatal period: future directions for immune research

So how does the above research about the immune etiology of OCD outside the perinatal period inform what we could or should be doing to research the immune etiology of perinatal OCD? Clearly, the work done so far on nonperinatal OCD has yielded no clear answers—cytokines may be lower or higher; autoantibodies may or may not exist; leukocyte subpopulations may or may not be altered. But combining this work with (1) knowledge of the most fruitful paths of immune research, those that are least susceptible to influences from other systems and (2) what we know from immune work in *other* psychiatric illnesses in the perinatal period may allow our field to avoid some of the less productive paths of research. Clearly research into circulating cytokines has yielded mixed results in both nonperinatal OCD and in perinatal depression—it may therefore make sense to skip this step, with the possible exception of IL-6, which has a strong body of evidence supporting its association with perinatal depression. Stimulating immune cells of subjects with perinatal OCD and measuring what cytokines they produce may be a promising avenue, as the few studies in this area in nonperinatal OCD have consistently found lower cytokine production from the stimulated cells of OCD subjects than from those of healthy controls. Examining autoantibodies, either those directed against the basal ganglia or those specific to autoimmune illnesses known to be elevated in other postpartum psychiatric disorders (such as antithyroid antibodies in postpartum psychosis (Wesseloo et al., 2018)), may also prove fruitful. Finally, examining leukocyte subpopulations may prove useful if we can concentrate on subsets known to be implicated in other immune illnesses, such as various T cell subsets (Osborne, Brar, & Klein, 2018). By taking advantage of the advances already made in these related areas, research into the immune system origins of perinatal OCD may thus be able to bypass some of the less useful pathways trod by their predecessors in these related fields to zero in on studies that may actually give us some answers about the etiology of an illness that affects many new mothers, impairs their ability to attach to their infants, and can often be misunderstood and mistaken for the more life-threatening postpartum psychosis.

References

Atmaca, M., Kilic, F., Koseoglu, F., & Ustundag, B. (2011). Neutrophils are decreased in obsessive-compulsive disorder: Preliminary investigation. *Psychiatry Investigation, 8*(4), 362–365. https://doi.org/10.4306/pi.2011.8.4.362.

Barber, Y., Toren, P., Achiron, A., Noy, S., Wolmer, L., Weizman, R., & Laor, N. (1996). T cell subsets in obsessive-compulsive disorder. *Neuropsychobiology, 34*(2), 63–66. https://doi.org/10.1159/000119293.

Bergink, V., Burgerhout, K. M., Weigelt, K., Pop, V. J., de Wit, H., Drexhage, R. C., … Drexhage, H. A. (2013). Immune system dysregulation in first-onset postpartum psychosis. *Biological Psychiatry, 73*(10), 1000–1007. https://doi.org/10.1016/j.biopsych.2012.11.006.

Bhattacharyya, S., Khanna, S., Chakrabarty, K., Mahadevan, A., Christopher, R., & Shankar, S. K. (2009). Anti-brain autoantibodies and altered excitatory neurotransmitters in obsessive-compulsive disorder. *Neuropsychopharmacology, 34*(12), 2489–2496. https://doi.org/10.1038/npp.2009.77.

Cappi, C., Brentani, H., Lima, L., Sanders, S. J., Zai, G., Diniz, B. J., … Fernandez, T. V. (2016). Whole-exome sequencing in obsessive-compulsive disorder identifies rare mutations in immunological and neurodevelopmental pathways. *Translational Psychiatry, 6*, e764. https://doi.org/10.1038/tp.2016.30.

Cappi, C., Muniz, R. K., Sampaio, A. S., Cordeiro, Q., Brentani, H., Palacios, S. A., … Hounie, A. G. (2012). Association study between functional polymorphisms in the TNF-alpha gene and obsessive-compulsive disorder. *Arquivos de Neuro-Psiquiatria, 70*(2), 87–90.

Carpenter, L. L., Heninger, G. R., McDougle, C. J., Tyrka, A. R., Epperson, C. N., & Price, L. H. (2002). Cerebrospinal fluid interleukin-6 in obsessive-compulsive disorder and trichotillomania. *Psychiatry Research, 112*(3), 257–262.

Chen, S. J., Liu, Y. L., & Sytwu, H. K. (2012). Immunologic regulation in pregnancy: From mechanism to therapeutic strategy for immunomodulation. *Clinical and Developmental Immunology, 2012*, 258391. https://doi.org/10.1155/2012/258391.

Cosco, T. D., Pillinger, T., Emam, H., Solmi, M., Budhdeo, S., Matthew Prina, A., … Carvalho, A. F. (2018). Immune aberrations in obsessive-compulsive disorder: A systematic review and meta-analysis. *Molecular Neurobiology*. https://doi.org/10.1007/s12035-018-1109-x.

Cox, C. J., Zuccolo, A. J., Edwards, E. V., Mascaro-Blanco, A., Alvarez, K., Stoner, J., … Cunningham, M. W. (2015). Antineuronal antibodies in a heterogeneous group of youth and young adults with tics and obsessive-compulsive disorder. *Journal of Child and Adolescent Psychopharmacology, 25*(1), 76–85. https://doi.org/10.1089/cap.2014.0048.

da Rocha, F. F., Correa, H., & Teixeira, A. L. (2008). Obsessive-compulsive disorder and immunology: A review. *Progress in Neuro-Psychopharmacology & Biological Psychiatry, 32*(5), 1139–1146. https://doi.org/10.1016/j.pnpbp.2007.12.026.

Dale, R. C. (2005). Post-streptococcal autoimmune disorders of the central nervous system. *Developmental Medicine and Child Neurology, 47*(11), 785–791. https://doi.org/10.1017/s0012162205001647.

Denys, D., Fluitman, S., Kavelaars, A., Heijnen, C., & Westenberg, H. G. (2006). Effects of paroxetine and venlafaxine on immune parameters in patients with obsessive compulsive disorder. *Psychoneuroendocrinology, 31*(3), 355–360. https://doi.org/10.1016/j.psyneuen.2005.08.016.

Dinn, W. M., Harris, C. L., McGonigal, K. M., & Raynard, R. C. (2001). Obsessive-compulsive disorder and immunocompetence. *International Journal of Psychiatry in Medicine, 31*(3), 311–320. https://doi.org/10.2190/f0ba-bn4f-61ka-ud99.

Fluitman, S., Denys, D., Vulink, N., Schutters, S., Heijnen, C., & Westenberg, H. (2010). Lipopolysaccharide-induced cytokine production in obsessive-compulsive disorder and generalized social anxiety disorder. *Psychiatry Research, 178*(2), 313–316. https://doi.org/10.1016/j.psychres.2009.05.008.

Fluitman, S. B., Denys, D. A., Heijnen, C. J., & Westenberg, H. G. (2010). Disgust affects TNF-alpha, IL-6 and noradrenalin levels in patients with obsessive-compulsive disorder. *Psychoneuroendocrinology, 35*(6), 906–911. https://doi.org/10.1016/j.psyneuen.2009.12.005.

Fontenelle, L. F., Barbosa, I. G., Luna, J. V., de Sousa, L. P., Abreu, M. N., & Teixeira, A. L. (2012). A cytokine study of adult patients with obsessive-compulsive disorder. *Comprehensive Psychiatry, 53*(6), 797–804. https://doi.org/10.1016/j.comppsych.2011.12.007.

Furtado, M., & Katzman, M. A. (2015). Neuroinflammatory pathways in anxiety, post-traumatic stress, and obsessive compulsive disorders. *Psychiatry Research, 229*(1–2), 37–48. https://doi.org/10.1016/j.psychres.2015.05.036.

Gause, C., Morris, C., Vernekar, S., Pardo-Villamizar, C., Grados, M. A., & Singer, H. S. (2009). Antineuronal antibodies in OCD: Comparisons in children with OCD-only, OCD+chronic tics and OCD+PANDAS. *Journal of Neuroimmunology, 214*(1–2), 118–124. https://doi.org/10.1016/j.jneuroim.2009.06.015.

Gray, S. M., & Bloch, M. H. (2012). Systematic review of proinflammatory cytokines in obsessive-compulsive disorder. *Current Psychiatry Reports, 14*(3), 220–228. https://doi.org/10.1007/s11920-012-0272-0.

Holtan, S. G., Chen, Y., Kaimal, R., Creedon, D. J., Enninga, E. A., Nevala, W. K., & Markovic, S. N. (2015). Growth modeling of the maternal cytokine milieu throughout normal pregnancy: Macrophage-derived chemokine decreases as inflammation/counterregulation increases. *International Journal of Immunology Research, 2015*, 952571. https://doi.org/10.1155/2015/952571.

Hounie, A. G., Cappi, C., Cordeiro, Q., Sampaio, A. S., Moraes, I., Rosario, M. C., … Miguel, E. C. (2008). TNF-alpha polymorphisms are associated with obsessive-compulsive disorder. *Neuroscience Letters, 442*(2), 86–90. https://doi.org/10.1016/j.neulet.2008.07.022.

Kant, R., Pasi, S., & Surolia, A. (2018). Auto-reactive Th17-cells trigger obsessive-compulsive-disorder like behavior in mice with experimental autoimmune encephalomyelitis. *Frontiers in Immunology, 9*, 2508. https://doi.org/10.3389/fimmu.2018.02508.

Konuk, N., Tekin, I. O., Ozturk, U., Atik, L., Atasoy, N., Bektas, S., & Erdogan, A. (2007). Plasma levels of tumor necrosis factor-alpha and interleukin-6 in obsessive compulsive disorder. *Mediators of Inflammation, 2007*, 65704. https://doi.org/10.1155/2007/65704.

Leff-Gelman, P., Mancilla-Herrera, I., Flores-Ramos, M., Cruz-Fuentes, C., Reyes-Grajeda, J. P., Garcia-Cuetara Mdel, P., … Pulido-Ascencio, D. E. (2016). The immune system and the role of inflammation in perinatal depression. *Neuroscience Bulletin, 32*(4), 398–420. https://doi.org/10.1007/s12264-016-0048-3.

Maes, M., Meltzer, H. Y., & Bosmans, E. (1994). Psychoimmune investigation in obsessive-compulsive disorder: Assays of plasma transferrin, IL-2 and IL-6 receptor, and IL-1 beta and IL-6 concentrations. *Neuropsychobiology, 30*(2–3), 57–60. https://doi.org/10.1159/000119136.

Maina, G., Albert, U., Bogetto, F., Borghese, C., Berro, A. C., Mutani, R., … Vigliani, M. C. (2009). Anti-brain antibodies in adult patients with obsessive-compulsive disorder. *Journal of Affective Disorders, 116*(3), 192–200. https://doi.org/10.1016/j.jad.2008.11.019.

Marazziti, D., Mucci, F., & Fontenelle, L. F. (2018). Immune system and obsessive-compulsive disorder. *Psychoneuroendocrinology, 93*, 39–44. https://doi.org/10.1016/j.psyneuen.2018.04.013.

Marazziti, D., Mungai, F., Masala, I., Baroni, S., Vivarelli, L., Ambrogi, F., … Dell'Osso, L. (2009). Normalisation of immune cell imbalance after pharmacological treatments of patients suffering from obsessive-compulsive disorder. *Journal of Psychopharmacology, 23*(5), 567–573. https://doi.org/10.1177/0269881108089605.

Marazziti, D., Presta, S., Pfanner, C., Gemignani, A., Rossi, A., Sbrana, S., … Cassano, G. B. (1999). Immunological alterations in adult obsessive-compulsive disorder. *Biological Psychiatry, 46*(6), 810–814.

Mataix-Cols, D., Frans, E., Perez-Vigil, A., Kuja-Halkola, R., Gromark, C., Isomura, K., … Larsson, H. (2018). A total-population multigenerational family clustering study of autoimmune diseases in obsessive-compulsive disorder and Tourette's/chronic tic disorders. *Molecular Psychiatry, 23*(7), 1652–1658. https://doi.org/10.1038/mp.2017.215.

Miman, O., Mutlu, E. A., Ozcan, O., Atambay, M., Karlidag, R., & Unal, S. (2010). Is there any role of Toxoplasma gondii in the etiology of obsessive-compulsive disorder? *Psychiatry Research, 177*(1–2), 263–265. https://doi.org/10.1016/j.psychres.2009.12.013.

Monteleone, P., Catapano, F., Fabrazzo, M., Tortorella, A., & Maj, M. (1998). Decreased blood levels of tumor necrosis factor-alpha in patients with obsessive-compulsive disorder. *Neuropsychobiology, 37*(4), 182–185. https://doi.org/10.1159/000026500.

Nicholson, T. R., Ferdinando, S., Krishnaiah, R. B., Anhoury, S., Lennox, B. R., Mataix-Cols, D., … Heyman, I. (2012). Prevalence of anti-basal ganglia antibodies in adult obsessive-compulsive disorder: Cross-sectional study. *British Journal of Psychiatry, 200*(5), 381–386. https://doi.org/10.1192/bjp.bp.111.092007.

Orlovska, S., Vestergaard, C. H., Bech, B. H., Nordentoft, M., Vestergaard, M., & Benros, M. E. (2017). Association of streptococcal throat infection with mental disorders: Testing key aspects of the PANDAS hypothesis in a nationwide study. *JAMA Psychiatry, 74*(7), 740–746. https://doi.org/10.1001/jamapsychiatry.2017.0995.

Osborne, L. M., Brar, A., & Klein, S. L. (2018). The role of Th17 cells in the pathophysiology of pregnancy and perinatal mood and anxiety disorders. *Brain, Behavior, and Immunity*. https://doi.org/10.1016/j.bbi.2018.11.015.

Pazos, M., Sperling, R. S., Moran, T. M., & Kraus, T. A. (2012). The influence of pregnancy on systemic immunity. *Immunologic Research, 54*(1–3), 254–261. https://doi.org/10.1007/s12026-012-8303-9.

Pearlman, D. M., Vora, H. S., Marquis, B. G., Najjar, S., & Dudley, L. A. (2014). Anti-basal ganglia antibodies in primary obsessive-compulsive disorder: Systematic review and meta-analysis. *British Journal of Psychiatry, 205*(1), 8–16. https://doi.org/10.1192/bjp.bp.113.137018.

Rao, N. P., Venkatasubramanian, G., Ravi, V., Kalmady, S., Cherian, A., & Yc, J. R. (2015). Plasma cytokine abnormalities in drug naive, comorbidity-free obsessive-compulsive disorder. *Psychiatry Research, 229*(3), 949–952. https://doi.org/10.1016/j.psychres.2015.07.009.

Ravindran, A. V., Griffiths, J., Merali, Z., & Anisman, H. (1999). Circulating lymphocyte subsets in obsessive compulsive disorder, major depression and normal controls. *Journal of Affective Disorders, 52*(1–3), 1–10.

Rodriguez, N., Morer, A., Gonzalez-Navarro, E. A., Gasso, P., Boloc, D., Serra-Pages, C., … Mas, S. (2017). Human-leukocyte antigen class II genes in early-onset obsessive-compulsive disorder. *World Journal of Biological Psychiatry*, 1–7. https://doi.org/10.1080/15622975.2017.1327669.

Rodriguez, N., Morer, A., Gonzalez-Navarro, E. A., Serra-Pages, C., Boloc, D., Torres, T., … Lazaro, L. (2017). Inflammatory dysregulation of monocytes in pediatric patients with obsessive-compulsive disorder. *Journal of Neuroinflammation, 14*(1), 261. https://doi.org/10.1186/s12974-017-1042-z.

Russell, E. J., Fawcett, J. M., & Mazmanian, D. (2013). Risk of obsessive-compulsive disorder in pregnant and postpartum women: A meta-analysis. *Journal of Clinical Psychiatry, 74*(4), 377–385. https://doi.org/10.4088/JCP.12r07917.

Sharma, T. R., Kline, D. B., Shreeve, D. F., & Hartman, D. W. (2011). Psychiatric comorbidities in patients with celiac disease: Is there any concrete biological association? *Asian Journal of Psychiatry, 4*(2), 150–151. https://doi.org/10.1016/j.ajp.2011.03.001.

Sherer, M. L., Posillico, C. K., & Schwarz, J. M. (2018). The psychoneuroimmunology of pregnancy. *Frontiers in Neuroendocrinology, 51*, 25–35. https://doi.org/10.1016/j.yfrne.2017.10.006.

Teixeira, A. L., Rodrigues, D. H., Marques, A. H., Miguel, E. C., & Fontenelle, L. F. (2014). Searching for the immune basis of obsessive-compulsive disorder. *Neuroimmunomodulation, 21*(2–3), 152–158. https://doi.org/10.1159/000356554.

Weizman, R., Laor, N., Barber, Y., Hermesh, H., Notti, I., Djaldetti, M., & Bessler, H. (1996). Cytokine production in obsessive-compulsive disorder. *Biological Psychiatry, 40*(9), 908–912. https://doi.org/10.1016/0006-3223(95)00520-x.

Wesseloo, R., Kamperman, A. M., Bergink, V., & Pop, V. J. M. (2018). Thyroid peroxidase antibodies during early gestation and the subsequent risk of first-onset postpartum depression: A prospective cohort study. *Journal of Affective Disorders, 225*, 399–403. https://doi.org/10.1016/j.jad.2017.08.058.

CHAPTER 10

Clinical phenotypes of postpartum psychosis

Katherine McEvoy
Perinatal Psychiatry, The Johns Hopkins University School of Medicine, Baltimore, Maryland, United States of America

Introduction

Postpartum psychosis (PPP) is a rare, but serious, psychiatric disorder that occurs in the early postnatal period. Classic features include psychosis (hallucinations, delusions, disorganized or bizarre behavior), mood symptoms (depression, mania, mixed symptoms), and cognitive dysfunction (disorientation, perplexity) (Sit, Rothschild, & Wisner, 2006). Some women with PPP appear delirious, confused, and disorganized. The majority of women will go on to experience manic episodes outside the perinatal period and receive a diagnosis of bipolar disorder, if not previously diagnosed, with a smaller subset having just puerperal episodes (Bergink, Ragson, and Wisner, 2016). PPP is a true psychiatric emergency, due to risk of suicide and infanticide, and requires immediate hospitalization for medical and psychiatric evaluation (Jones, Chandra, Dazzan, & Howard, 2014; Sit et al., 2006).

Terminology

Some consider PPP a distinct psychiatric disorder with classic features (as above), distinguished from an episode of mania or depression with psychosis by the characteristic altered mental status. Others consider any psychotic illness occurring postpartum to fall under the umbrella of PPP. This would include postpartum depression (PPD) with psychosis, mania with psychosis, or an exacerbation of schizophrenia. Having a diagnosis of schizophrenia or major depressive disorder (MDD) with psychosis increases a woman's chance of having classic PPP, but such women may also develop an exacerbation of their usual illness (Sit et al., 2006). These differences in terminology lead to important scientific differences as well; some studies have included patients with MDD with psychosis in their sample. If the

Biomarkers of Postpartum Psychiatric Disorders
ISBN 978-0-12-815508-0
https://doi.org/10.1016/B978-0-12-815508-0.00010-2

disorder is truly a distinct entity, then such samples are not homogeneous, thus limiting extrapolation of findings. Leading researchers in PPP have commented that this confusion with nosology and potential misclassification of cases has likely hindered research in the area (Jones et al., 2014). With those caveats in mind, for the purposes of this book chapter, we will consider PPP as a distinct psychiatric entity.

Epidemiology

PPP is a rare disorder, occurring in 0.5 (Florio, Munk-Olsen, & Bergink, 2016) to 2.6 (VanderKruik et al., 2017) per 1000 women and is more likely to occur in primiparous women (Sit et al., 2006). This risk increases substantially, however, to 29%, for those women who have had a previous episode of PPP (Wesseloo et al., 2016), and women with bipolar disorder are also at significantly increased risk, occurring in up to 20% of cases (Jones et al., 2014). It is well established that the postpartum period poses the highest lifetime risk for a woman of hospitalization for psychosis (Kendell, Chalmers, & Platz, 1987; Terp & Mortensen, 1998; Tschinkel, Harris, Le Noury, & Healy, 2007), and a woman is 23 times more likely to experience a first-onset affective psychosis in the first 4 weeks postpartum than any other time in her life (Munk-Olsen, Laursen, Pedersen, Mors, & Mortensen, 2006).

Risk factors

Box 10.1 lists the established risk factors. The single strongest risk factor is a preexisting diagnosis of bipolar disorder, with 20%–30% of these women going on to experience PPP (Di Florio et al., 2013). Despite this, the majority of women with PPP do *not* have a preexisting diagnosis of bipolar disorder (Blackmore et al., 2013); because the majority of women with a first-time PPP will go on to be diagnosed with bipolar disorder, this makes the postpartum a time of increased risk for a first episode of bipolar illness. Of the women who present with PPP as their first psychiatric episode, 50%–80% of them will go on to have nonpuerperal episodes, mostly bipolar disorder (Chaudron & Pies, 2003; Nager, Szulkin, Johansson, Johansson, & Sundquist, 2013). It is presently not clear if obstetric complications are risk factors for PPP, as the current literature is mixed. Some studies suggest an association (Bergink, Laursen, et al., 2015; Blackmore et al., 2006, 2013) while others do not (Kumar, Marks, Platz, &

Box 10.1 Risk factors for PPP (Jones and Craddock, 2001, 2002; Kendell et al., 1987)

- Preexisting bipolar disorder
- History of postpartum psychosis
- Family history of postpartum psychosis
- Family history of bipolar disorder
- Schizophrenia or schizoaffective disorder
- Discontinuation of psychiatric medications for pregnancy (Viguera et al., 2000, 2007)
- Primiparity
- Vulnerability to hormone shifts
- Immune dysfunction

Yoshida, 1995; Meltzer-Brody et al., 2017; Munk-Olsen et al., 2006). Sleep disturbance has not been studied specifically in relation to PPP, but it is an established risk factor for mania in bipolar disorder, so it would not be surprising if it were also implicated in PPP. Additionally, long labors or overnight deliveries are associated with PPP (Lewis, Foster, & Jones, 2016; Sharma, Smith, & Khan, 2004). A recent study found that women with bipolar disorder who report decreased sleep as a trigger for manic episodes are at increased risk of PPP related to sleep deprivation (Lewis et al., 2018). Primiparity (Jones et al., 2014) has been shown to increase risk even when controlling for the fact that women with PPP are less likely to have a second baby. It is not known if this association is related to biological or psychosocial factors, or both. On the other hand, stressful life events do *not predict* PPP (Brockington, Martin, Brown, Goldberg, & Margison, 1990; Marks, Wieck, Checkley & Kumar, 1992; McNeil, 1988), as they do with episodes of PPD. Genetic, immune, and hormonal risk factors will be discussed in subsequent chapters.

Clinical presentation

Patients present with psychotic symptoms (delusions, hallucinations, thought disorder, grossly disorganized or bizarre behavior), which is most often accompanied by a disturbance in mood—depressed, manic, or mixed (Jones et al., 2014). The onset is usually rapid, occurring in the first 2–4 weeks postpartum, with up to 50% occurring as early 2–3 days postpartum (Jones et al., 2014). Clinical symptoms vary widely among patients, and symptoms can shift rapidly within individual patients, reflected in the term "the kaleidoscopic picture of postpartum psychosis" (Bergink et al., 2016). Early symptoms may include decreased need for sleep or inability to sleep, feeling elated or excited, being more talkative, and having more energy (Heron, McGuinness, Blackmore, Craddock, & Jones, 2008). Families may notice a change in behavior first and bring the woman to clinical attention (Bergink et al., 2016). The women themselves often lose insight, especially if there is a predominance of manic symptoms, and their primary concern may be related to a delusion, such as their partner trying to harm the baby. Chief complaints can include confusion, agitation, inability to sleep, bizarre or disorganized behavior, delusional beliefs, and perceptual disturbances. Suicidal or infanticidal ideas may rarely occur. A cohort study of 130 consecutive cases of PPP used latent cluster analysis to describe three separate symptom profiles (Kamperman, Veldman-Hoek, Wesseloo, Robertson Blackmore, & Bergink, 2017):

1. Cases characterized by mania and/or agitation, with irritability being much more common than euphoria (34%)
2. Cases characterized by depression and/or anxiety (41%)
3. Cases with an atypical profile (25%)

It is interesting to note that all cases with cognitive dysfunction, such as disorientation and clouding of consciousness, occurred in the "atypical" group, which amounted to only 25% of the sample. This is a significant finding, as prior to this study, cognitive dysfunction was considered one of the hallmarks of PPP. It is also possible, however, that some of the women labeled as PPP with depressive phenotype were actually cases of MDD with psychosis. Also of note in this cohort, 72% had abnormal thought content, most commonly persecutory delusions and delusions of reference, 39% reported auditory hallucinations, one in three women reported visual hallucinations, one-quarter had disorganized behavior, and 5% had catatonic symptoms. It is noteworthy that visual hallucinations only rarely occur in bipolar episodes outside the postpartum period.

In general, the majority of delusions relate to the infant (Chandra, Bhargavaraman, Raghunandan, & Shaligram, 2006) or childbirth (Bergink et al., 2016). Delusions of "altruistic homicide" can also occur and so must be explored during the initial evaluation (Resnick, 1969). Risks of suicide or infanticide may also be related to auditory command hallucinations. First-rank symptoms, such as thought insertion or withdrawal and passivity delusions, rarely occur (Bergink et al., 2016).

Investigations into maternal deaths in the United Kingdom have shown that suicide risk is increased 70-fold in the first year after childbirth, and that suicide is the leading cause of maternal death in first year after childbirth in countries that track it (CEMD, 2001). Approximately 4% of women with a diagnosis of PPP will die by suicide (Pfuhlmann, Stoeber, & Beckmann, 2002) and most often by violent means, which is in contrast to less violent means of suicide used by women outside of the postnatal time-period (CEMD, 2001).

Infanticidal ideas and behavior are also a feature of PPP, though they are not common (Sit et al., 2006). In a sample of women hospitalized on a mother–baby unit, only 9% had thoughts of harming their babies, while up to 35% had delusions related to the baby (Kumar et al., 1995). Rates of infanticidal ideas seem to differ by region and perhaps culture, with a 9% rate in a Dutch sample, 9% in a British sample, 33% in an American sample, and almost 50% in an Indian sample. Given the potential dire consequences, the examiner should carefully question the woman about such ideas (Kamperman et al., 2017). In addition to thoughts of directly harming the infant, those women with cognitive dysfunction or grossly disorganized behavior are at risk of indirectly harming the baby through neglect or unsafe actions (Kumar et al., 1995). The overall rate of infanticide is 4% (Spinelli, 2009).

Boxes 10.2 and 10.3 show examples of two different case presentations.

Assessment and diagnosis

Once there is a concern for PPP, the woman should have an expedited psychiatric and medical evaluation, preferably as an inpatient (Sit et al., 2006). Ideally, this should take place in a monitored setting where the baby can remain with the mother, but under close supervision, though there are no overnight mother–baby units presently in the United States. PPP is a clinical diagnosis based on the relevant symptoms and signs in a woman with risk factors but requires a careful workup to exclude organic causes of psychosis (Sit et al., 2006). Boxes 10.4 and 10.5 outline suggested screening questions and workup.

DSM-5 does not recognize PPP as a distinct disorder but allows for the use of a "peripartum onset" specifier with another mental illness diagnosis, such as "MDD with psychosis, with peripartum onset" or "bipolar disorder, current episode manic, with psychotic features, with peripartum onset" (APA, 2013; DSM5).

Box 10.2 Case 1: Postpartum psychosis with manic features

Ms. B was an emergency room physician with no prior psychiatric history who delivered her first baby by vaginal delivery 10 days before her psychiatric admission. She had a fever 24 h after delivery and was found to have a group B strep urinary tract infection (UTI), treated successfully with antibiotics. While in the hospital she was observed to be anxious and vigilant about the baby's well-being and slept very little. After discharge on day 3 following delivery, she continued to be hypervigilant at home and the family reported that she did not sleep for the 7 days prior to presentation. She did not trust other family members with the care of the baby and kept the infant in her care 24/7. Two days prior to presentation, she became suspicious of other family members and accused them of putting bleach in the baby's bottle and on the baby's genitals. The night before presentation, her husband found her outside with the baby in the snow, neither of them appropriately dressed, disposing of all the cleaning agents in the house. She remained suspicious of her husband and refused to go to the hospital. The family later called the police in order to escort her to the hospital. In the emergency room she had pressured speech and accused the staff of conspiring with her family to remove her baby and harm her.

Box 10.3 Case 2: Postpartum psychosis with depressed features

Ms. M was a 19-year-old woman with a history of childhood trauma and domestic violence, who delivered a preterm infant vaginally at 34 weeks without the support of the baby's father. The baby spent 3 days in the NICU for prematurity and the patient remained with the baby. Although constantly present, she spent most of her time on her phone or watching TV and rarely asked questions about the baby's well-being. She was noted to be tearful on numerous occasions, but this was felt to be appropriate given the context and she was not referred for mental health treatment. She lived with her mother and she returned there with the baby following discharge. Her mother reported that once they returned home the patient withdrew more and appeared to be moving around very slowly and barely spoke, which was unlike her. She was unable to care for the baby at all and stopped eating and drinking. She made statements like "I am not a mother," "my baby is dead," "I am dead." On the day of presentation, she was found toileting in one of the bedrooms and was disoriented. Her family brought her to the emergency room and she appeared frightened and confused on initial exam. She had significant speech latency and was disoriented to time and place. She denied having a baby.

Box 10.4 Important screening questions for patient and family

- Has the patient had prior psychiatric treatment?
- If so, was it for depression, mania, or both?
- Has the patient ever displayed odd or bizarre behavior?
- Is there a family history of bipolar disorder or postpartum psychosis?
- Does the patient abuse drugs or alcohol?
- Does the patient have any medical conditions?
- Does the patient have any thoughts of harming herself or her infant? It is important to ask this in a sensitive manner, as women may conceal their ideas to hurt their infant. A good way to approach it is to normalize it by saying "a lot of women experience this …, have you had something similar?"

Box 10.5 Important assessment tools

- Comprehensive physical exam, including a neurological exam
- Complete blood count
- Complete metabolic panel
- Urinalysis
- Urine toxicology screen
- Thyroid-stimulating hormone, free T4, and thyroid peroxidase antibodies
- Ammonia level
- B12 and folate
- If warranted, brain imaging and testing for limbic encephalitis

Differential diagnosis

Baby blues. This syndrome is considered normal and is found in up to 80% of postpartum women. It usually consists of mild emotional lability and lasts no more than 10 days. Psychotic symptoms of any sort, functional impairment, as well as suicidal or homicidal thoughts should not be present.

PPD with psychosis. The patient will usually have had prior episodes of depression, and during the episode there should not be any cognitive dysfunction or grossly disorganized behavior. Symptoms should not fluctuate as intensely.

Postpartum OCD. This may present with intrusive and unwanted thoughts of the woman harming the infant, which need to be distinguished from psychotic delusions or direct desires or intentions to harm the infant. Women with OCD are usually horrified by these thoughts and have no desire to carry them out. They retain insight and are not considered at risk of harming their infants (Sit et al., 2006).

Generalized anxiety disorder. This is common in the postpartum period and can result in increased activity and decreased sleep, which can simulate mania. The patient will retain insight and there should be no psychotic symptoms.

Schizophrenia or schizoaffective disorder. There should be a prior history of episodes of these disorders. Exacerbations of these disorders are usually similar over time, usually with the same or similar delusions. There may also be first-rank symptoms present, which are uncharacteristic of PPP.

Substance abuse disorders. Psychotic symptoms or cognitive dysfunction could result from either intoxication or withdrawal.

Infection. Examples include endometritis, UTI, or wound infection.

Cerebrovascular events. Hypertensive emergency; stroke, hemorrhage. Cranial imaging is indicated if there are any symptoms or signs of these.

Flares of autoimmune disorders. Lupus or sarcoidosis.

Autoimmune thyroiditis. Occurs in 5%–7% of women in the postpartum period and 20% of women with PPP (Bergink et al., 2011; Muller, Drexhage, & Berghout, 2001).

Autoimmune encephalitis. Neurological symptoms such as seizures, drowsiness, and extrapyramidal signs should make one suspicious for anti-NMDA receptor encephalitis. Peripheral antibodies can be falsely positive, so CSF testing is preferred if warranted (Bergink, Armangue, et al., 2015).

Metabolic abnormalities. Hypo- or hypernatremia, hypo- or hypercalcemia, hypoglycemia.

Medication reaction. E.g., steroid-induced mania.

Sheehan syndrome. This is infarction of the pituitary gland, resulting in acute hypopituitarism following severe postpartum hemorrhage, more common in developing countries (Reddy, Nahar, Thippeswamy, & Kumar, 2017).

Inborn errors of metabolism. These may manifest for the first time postpartum, which is a significant catabolic state. Ammonia will be high in urea cycle disorders (Bergink et al., 2016).

Screening and prevention

There are no validated screening tools for PPP, but all women should be screened (ideally during pregnancy) for a history of prior mental illness—especially prior manias, psychosis, or a previous PPP. A family history of bipolar disorder, PPP, or any other type of psychosis should also be ascertained. Patients with a positive personal or family history should be educated about the risk of becoming ill postpartum and what symptoms and signs to look out for. These patients should also receive additional monitoring postpartum, and those with a prior PPP may benefit from prophylactic treatment (Sit et al., 2006). In a small study looking at 29 women with a prior history of PPP (without nonpuerperal episodes), 44% of women who declined prophylactic postpartum treatment with lithium relapsed, whereas none of the women who underwent prophylactic treatment did (Bergink et al., 2012). Patients with bipolar disorder should be educated about the substantially higher risk of PPP in those who discontinue medications for pregnancy (Viguera et al., 2000). All women should be asked about symptoms of PPP at the 6 week postpartum visit.

References

American Psychiatric Association, Diagnostic and statistical manual of mental disorders, (DSM-5). (5th ed). (2013). Arlington: American Psychiatric Association.

Bergink, V., Armangue, T., Titulaer, M. J., Markx, S., Dalmau, J., & Kushner, S. A. (2015). Autoimmune encephalitis in postpartum psychosis. *The American Journal of Psychiatry, 172*(9), 901–908. https://doi.org/10.1176/appi.ajp.2015.14101332.

Bergink, V., Bouvy, P. F., Vervoort, J. S., Koorengevel, K. M., Steegers, E. A., & Kushner, S. A. (2012). Prevention of postpartum psychosis and mania in women at high risk. *The American Journal of Psychiatry, 169*(6), 609–615.

Bergink, V., Kushner, S. A., Pop, V., Kuijpens, H., Lambregtse-van den Berg, M. P., Drexhage, R. C., … Drexhage, H. A. (2011). Prevalence of autoimmune thyroid dysfunction in postpartum psychosis. *The British Journal of Psychiatry : The Journal of Mental Science, 198*(4), 264–268. https://doi.org/10.1192/bjp.bp.110.082990.

Bergink, V., Laursen, T. M., Johannsen, B. M., Kushner, S. A., Meltzer-Brody, S., & Munk-Olsen, T. (2015a). Pre-eclampsia and first-onset postpartum psychiatric episodes: A Danish population-based cohort study. *Psychological Medicine, 45*(16), 3481–3489.

Bergink, V., Rasgon, N., & Wisner, K. L. (2016). Postpartum psychosis: Madness, mania, and melancholia in motherhood. *The American Journal of Psychiatry, 173*(12), 1179–1188.

Blackmore, E. R., Jones, I., Doshi, M., Haque, S., Holder, R., Brockington, I., & Craddock, N. (2006). Obstetric variables associated with bipolar affective puerperal psychosis. *British Journal of Psychiatry, 188*, 32–36.

Blackmore, E. R., Rubinow, D. R., O'Connor, T. G., Liu, X., Tang, W., Craddock, N., & Jones, I. (2013). Reproductive outcomes and risk of subsequent illness in women diagnosed with postpartum psychosis. *Bipolar Disorders, 15*(4), 394–404.

Brockington, I. F., Martin, C., Brown, G. W., Goldberg, D., & Margison, F. (1990). Stress and puerperal psychosis. *British Journal of Psychiatry, 157*, 331–334.

CEMD. (2001). *Confidential inquiries into maternal deaths: Why mothers die* (pp. 1997–1999). London: Royal College of Obstetricians and Gynaecologists.

Chandra, P. S., Bhargavaraman, R. P., Raghunandan, V. N., & Shaligram, D. (2006). Delusions related to infant and their association with mother-infant interactions in postpartum psychotic disorders. *Archives of Women's Mental Health, 9*(5), 285–288.

Chaudron, L. H., & Pies, R. W. (2003). The relationship between postpartum psychosis and bipolar disorder: A review. *Journal of Clinical Psychiatry, 64*, 1284–1292.

Di Florio, A., Forty, L., Gordon-Smith, K., Heron, J., Jones, L., Craddock, N., & Jones, I. (2013). Perinatal episodes across the mood disorder spectrum. *JAMA Psychiatry, 70*(2), 168–175.

Florio, A. D., Munk-Olsen, T., & Bergink, V. (2016). The birth of a psychiatric orphan disorder: Postpartum psychosis. *The Lancet. Psychiatry, 3*(6), 502, 500366(16)30037-2.

Heron, J., McGuinness, M., Blackmore, E. R., Craddock, N., & Jones, I. (2008). Early postpartum symptoms in puerperal psychosis. *BJOG: An International Journal of Obstetrics and Gynaecology, 115*(3), 348–353.

Jones, I., Chandra, P. S., Dazzan, P., & Howard, L. M. (2014). Bipolar disorder, affective psychosis, and schizophrenia in pregnancy and the post-partum period. *Lancet, 384*(9956), 1789. https://doi.org/10.1016/S0140-6736(14)61278-2.

Jones, I., & Craddock, N. (2001). Familiality of the puerperal trigger in bipolar disorder: Results of a family study. *American Journal of Psychiatry, 158*(6), 913–917.

Jones, I., & Craddock, N. (2002). Do puerperal psychotic episodes identify a more familial subtype of bipolar disorder? Results of a family history study. *Psychiatric Genetics, 12*(3), 177–180.

Kamperman, A. M., Veldman-Hoek, M. J., Wesseloo, R., Robertson Blackmore, E., & Bergink, V. (2017). Phenotypical characteristics of postpartum psychosis: A clinical cohort study. *Bipolar Disorders, 19*(6), 450–457.

Kendell, R. E., Chalmers, J. C., & Platz, C. (1987). Epidemiology of puerperal psychoses. *British Journal of Psychiatry, 150*(MAY), 662–673.

Kumar, R., Marks, M., Platz, C., & Yoshida, K. (1995). Clinical survey of a psychiatric mother and baby unit: Characteristics of 100 consecutive admissions. *Journal of Affective Disorders, 33*(1), 11–22.

Lewis, K. J. S., Di Florio, A., Forty, L., Gordon-Smith, K., Perry, A., Craddock, N., ... Jones, I. (2018). Mania triggered by sleep loss and risk of postpartum psychosis in women with bipolar disorder. *Journal of Affective Disorders, 225*, 624–629.

Lewis, K. J., Foster, R. G., & Jones, I. R. (2016). Is sleep disruption a trigger for postpartum psychosis? *The British Journal of Psychiatry: The Journal of Mental Science, 208*(5), 409–411.

Marks, M. N., Wieck, A., Checkley, S. A., & Kumar, R. (1992). Contribution of psychological and social factors to psychotic and non-psychotic relapse after childbirth in women with previous histories of affective disorder. *Journal of Affective Disorders, 24*(4), 253–263.

McNeil, T. F. (1988). A prospective study of postpartum psychoses in a high-risk group. 4. relationship to life situation and experience of pregnancy. *Acta Psychiatrica Scandinavica, 77*(6), 645–653.

Meltzer-Brody, S., Maegbaek, M. L., Medland, S. E., Miller, W. C., Sullivan, P., & Munk-Olsen, T. (2017). Obstetrical, pregnancy and socio-economic predictors for new-onset severe postpartum psychiatric disorders in primiparous women. *Psychological Medicine, 47*(8), 1427–1441.

Muller, A. F., Drexhage, H. A., & Berghout, A. (2001). Postpartum thyroiditis and autoimmune thyroiditis in women of childbearing age: Recent insights and consequences for antenatal and postnatal care. *Endocrine Reviews, 22*(5), 605–630.

Munk-Olsen, T., Laursen, T. M., Pedersen, C. B., Mors, O., & Mortensen, P. B. (2006). New parents and mental disorders: A population-based register study. *Journal of the American Medical Association, 296*(21), 2582–2589.

Nager, A., Szulkin, R., Johansson, S. E., Johansson, L. M., & Sundquist, K. (2013). High lifelong relapse rate of psychiatric disorders among women with postpartum psychosis. *Nordic Journal of Psychiatry, 67*(1), 53–58.

Pfuhlmann, B., Stoeber, G., & Beckmann, H. (2002). Postpartum psychoses: Prognosis, risk factors, and treatment. *Current Psychiatry Reports, 4*(3), 185–190.

Reddy, M. S., Nahar, A., Thippeswamy, H., & Kumar, C. S. (2017). Psychosis as a late manifestation of sheehan's syndrome. *Asian Journal of Psychiatry, 25*, 228–230.

Resnick, P. J. (1969). Child murder by parents: A psychiatric review of filicide. *The American Journal of Psychiatry, 126*(3), 325–334.

Sharma, V., Smith, A., & Khan, M. (2004). The relationship between duration of labour, time of delivery, and puerperal psychosis. *Journal of Affective Disorders, 83*(2–3), 215–220.

Sit, D., Rothschild, A. J., & Wisner, K. L. (2006). A review of postpartum psychosis. *Journal of Women's Health (2002), 15*(4), 352–368. https://doi.org/10.1089/jwh.2006.15.352.

Spinelli, M. G. (2009). Postpartum psychosis: Detection of risk and management. *The American Journal of Psychiatry, 166*(4), 405–408.

Terp, I. M., & Mortensen, P. B. (1998). Post-partum psychoses: Clinical diagnoses and relative risk of admission after parturition. *British Journal of Psychiatry, 172*, 521–526.

Tschinkel, S., Harris, M., Le Noury, J., & Healy, D. (2007). Postpartum psychosis: Two cohorts compared, 1875–1924 and 1994–2005. *Psychological Medicine, 37*(4), 529–536.

VanderKruik, R., Barreix, M., Chou, D., Allen, T., Say, L., Cohen, L. S., ... von Dadelszen, P. (2017). The global prevalence of postpartum psychosis: A systematic review. *BMC Psychiatry, 17*(1), 272-017-1427-7.

Viguera, A. C., Nonacs, R., Cohen, L. S., Tondo, L., Murray, A., & Baldessarini, R. J. (2000). Risk of recurrence of bipolar disorder in pregnant and nonpregnant women after discontinuing lithium maintenance. *The American Journal of Psychiatry, 157*(2), 179–184.

Viguera, A. C., Whitfield, T., Baldessarini, R. J., Newport, D. J., Stowe, Z., Reminick, A., ... Cohen, L. S. (2007). Risk of recurrence in women with bipolar disorder during pregnancy: Prospective study of mood stabilizer discontinuation. *The American Journal of Psychiatry, 164*(12), 1817–1824. quiz 1923.

Wesseloo, R., Kamperman, A. M., Munk-Olsen, T., Pop, V. J., Kushner, S. A., & Bergink, V. (2016). Risk of postpartum relapse in bipolar disorder and postpartum psychosis: A systematic review and meta-analysis. *The American Journal of Psychiatry, 173*(2), 117–127.

CHAPTER 11

Genetic basis for postpartum psychosis

Arianna Di Florio[1,2]
[1]Cardiff University, Cardiff, United Kingdom; [2]University of North Carolina at Chapel Hill, Chapel Hill, NC, United States of America

Introduction

The term "postpartum psychosis" is commonly used to describe the acute onset of psychotic or manic symptoms in the first few weeks after delivery (Brockington et al., 1981; Dean & Kendell, 1981; Di Florio, Smith, & Jones, 2013; Heron, Robertson Blackmore, McGuinness, Craddock, & Jones, 2007). It is considered a rare disorder, with incidence ranging between 0.25 and 0.50 per 1000 deliveries (Florio, Munk-Olsen, & Bergink, 2016). Despite the low incidence, the first 4 weeks after childbirth are the period of highest risk of first episode affective psychosis in women (Munk-Olsen, Laursen, Pedersen, Mors, & Mortensen, 2006).

Postpartum psychosis is not officially recognized by the current diagnostic systems, the *Diagnostic and Statistical Manual of Mental Disorders* (DSM) curated by the American Psychiatric Association (Diagnostic and Statistical Manual of Mental Disorders | DSM Library, n.d.) and the World Health Organization's International Classification of Diseases (ICD) (ICD-10 Version:2010, n.d.). Rather, it is diagnosed with the general mania or psychotic depression categories of the mood disorder rubric. The relationship to childbirth is flagged by the use of a perinatal onset specifier, which covers pregnancy and 4 weeks postpartum in DSM 5 and "about" 6 weeks postpartum in ICD-11.

The etiopathology of postpartum psychosis is still unknown. Confusion about terminology and classification has hindered research in this area (Florio et al., 2016), though there is robust evidence of a link between childbirth and maternal mental disorders. Evidence suggests that the association is specific or stronger for bipolar disorder (vs. schizophrenia or major depression) (Munk Olsen et al., 2009, 2006), for manic episodes (vs. nonpsychotic bipolar depression) (Di Florio, Forty, et al., 2013), and for childbirth (vs. miscarriage or termination) (Di Florio et al., 2015). The

Biomarkers of Postpartum Psychiatric Disorders
ISBN 978-0-12-815508-0
https://doi.org/10.1016/B978-0-12-815508-0.00011-4

significantly increased risk of an episode of bipolar illness in the first few weeks after childbirth is reported both in primiparous women without a previous psychiatric history (Munk-Olsen et al., 2006) and in those with a history of bipolar disorder (Munk-Olsen et al., 2009). Interestingly, the presentation of any psychiatric episode in the immediate postpartum period puts women at significantly increased risk of developing bipolar affective disorder during a 15-year follow-up period (Munk-Olsen, Laursen, Meltzer-Brody, Mortensen, & Jones, 2012).

The abrupt onset of severe affective psychosis in women often without previous history of mental disorders (Florio et al., 2018; Munk-Olsen et al., 2006), the lack of association with socioeconomic and environmental factors (Brockington, Martin, Brown, Goldberg, & Margison, 1990; Dowlatshahi & Paykel, 1990; Marks, Wieck, Checkley, & Kumar, 1991; Perry et al., 2016), and similar incidence rates across countries and cultures (Nakigudde, Ehnvall, Mirembe, Musisi, & Airaksinen, 2013; Posmontier & Fisher, 2014; Thippeswamy, Dahale, Desai, & Chandra, 2015; VanderKruik et al., 2017) have suggested that biological factors may play an important role in the susceptibility to postpartum psychosis.

Genetic biomarkers of postpartum psychosis have not been discovered yet. Therefore, our discussion here is limited to clinical and epidemiological evidence on the familiality of postpartum psychosis. We will then extend our focus to molecular genetic studies of bipolar disorder, as it has been hypothesized that the majority of episodes of postpartum psychosis are manifestations of a bipolar disorder diathesis with a puerperal trigger. We will conclude by suggesting that postpartum psychosis represents a great opportunity for genetic biomarker discovery.

Clinical and epidemiological studies

Clinical and epidemiological studies on the familiality of postpartum psychosis have mainly focused on two key questions: (1) Is there a familial association between postpartum psychosis and other mental disorders? This question can be subdivided in two parts: Are family members of women with postpartum psychosis at higher risk of mental disorders than the general population? Are women with a family history of mental disorders at higher risk of postpartum psychosis compared to the general population? and (2) Is there a familial component that is specific to the puerperal triggering?

Familial association between postpartum psychosis and other mental disorders

There is evidence that family members of women with postpartum psychosis have higher rates of mental disorders than the general population, but evidence is not univocal. Different definitions of postpartum psychosis and family history of mental disorders have been used, and comparisons across studies are difficult. Depending on the definitions used, 30%–65% of women with postpartum psychosis have family members affected by psychiatric disorders, with morbidity risks for first-degree relatives in the range of 10%–50% (Dean, Williams, & Brockington, 1989; Huhn & Drenk, 1973; Jones & Craddock, 2001; Ian; Jones & Craddock, 2002; Kadrmas, Winokur, & Crowe, 1979; Osterman, 1963; Platz & Kendell, 1988; Protheroe, 1969; Schöpf, Bryois, Jonquiére, & Scharfetter, 1985; Schöpf & Rust, 1994; Seager, 1960; Thuwe, 1974; Whalley, Roberts, Wentzel, & Wright, 1982). The majority of studies, however, used a family history approach (one informant, usually the proband, and/or case notes) rather than a family study approach (direct ascertainment of family members). Moreover, they did not include matched controls and, therefore, were not able to provide an estimate of the risk relative to the general population.

Several studies have suggested an overlap in the familiality between postpartum psychosis and bipolar affective disorder. Moreover, rates of affective disorders in first-degree relatives of women with postpartum psychosis are significantly higher than those in relatives of parous women with bipolar disorder without postpartum psychosis (Dean et al., 1989; Jones & Craddock, 2001). These findings suggest that postpartum psychosis may represent a marker for a more familial subtype of bipolar disorder.

A recent epidemiological, register-based study on the entire Danish population has investigated the familial risk of psychiatric episodes within 6 months postpartum (i.e., whether family history predicts the risk of any psychiatric episode requiring hospital or outpatient treatment within 6 months after delivery) (Bauer et al., 2018). The study found that the risk of psychiatric episodes was significantly elevated in women with a first-degree relative with psychiatric disorders compared to controls. Consistent with previous observations on women with postpartum psychosis (Dean et al., 1989; Jones & Craddock, 2002), the strongest association was reported for bipolar disorder. Women without previous psychiatric history had an over threefold increased risk of any postpartum episode if they had a first-degree relative with bipolar disorder (hazard ratio: 3.61; 95% confidence intervals: 2.09–6.23). This study, however, included all psychiatric episodes occurring within 6 months postpartum, not only postpartum psychosis.

Familial association of the puerperal trigger

There is a paucity of studies investigating the specificity of the postpartum triggering of severe episodes. Jones and Craddock examined this question in a family study with multiple members affected by bipolar disorder and/or postpartum psychosis (Jones & Craddock, 2001). They found that rates of postpartum psychosis were 570 in 1000 deliveries in women with a family history of postpartum psychosis, versus 260 in 1000 deliveries for women with bipolar disorder without a family history of postpartum psychosis. Other studies have investigated the specificity of the familiality of the puerperal trigger, but included any affective disorder, not just bipolar disorder (Platz & Kendell, 1988; Schöpf et al., 1985; Schöpf & Rust, 1994; Thuwe, 1974). Results therefore are not comparable. Based on Jones and Craddock's findings, we conclude that there seems to be a specific familial component to the susceptibility to the puerperal triggering.

Taken together, these studies provide robust evidence for the familiality of postpartum psychosis but do not explore the heritability of postpartum psychosis. Genetic factors are only one explanation for familiality. Other factors such as shared lifestyle or environment can explain why a trait runs in families. Not all familial traits are therefore heritable. Traits are heritable only if the similarity in the phenotype arises from shared genotypes. There are no twin or adoption studies of puerperal psychosis to assess the importance of genetic factors. However, reports of identical twin pairs concordant for puerperal psychosis (I. Jones & Craddock, 2001), as well as familial clustering of puerperal psychosis associated with close consanguinity (Craddock et al., 1994), raise the possibility of genetic factors contributing to this familiality.

Molecular genetic studies of bipolar disorder may also provide important clues into the genetic liability to postpartum psychosis. Such studies have in fact unequivocally established the role of many genetic variants of small effect in the susceptibility to common psychiatric disorders (International Schizophrenia Consortium et al., 2009; Ripke et al., 2014; Ruderfer et al., 2014; Wray et al., 2014). Bipolar disorder genome-wide association studies are discussed in the next section, as they may represent an important anchor point in the discovery of genetic markers for postpartum psychosis.

Molecular genetic studies

Research into genetic biomarkers of postpartum psychosis is limited to candidate gene studies (Coyle, Jones, Robertson, Lendon, & Craddock, 2000; Feng et al., 2001; Jones et al., 2000; Middle et al., 2003) and a genome-wide linkage study (Jones et al., 2007). The latter has implicated an area on chromosome 16p13 that may harbor susceptibility genes for puerperal psychosis, but follow-up studies have never been conducted to explore this region further.

Genome-wide association studies have provided important and robust insight into the biology and nosology of psychiatric disorders, including bipolar disorder. The largest genome-wide association study of bipolar disorder (Bipolar Disorder Working Group, 2019) has identified 30 genome-wide significant loci involved in the susceptibility to bipolar disorder. Top hits have implicated genes that had not been hypothesized as candidate genes for bipolar disorder before, such as those encoding for fatty acid desaturases (FADS) or the calcium channel α1 subunit (CACNA1). A second major discovery has been that noncoding genetic variants, such as regulatory genomic elements, play an important role in the liability to neuropsychiatric disorders, including bipolar disorder (Boyle, Li, & Pritchard, 2017; Hou et al., 2016; Starnawska et al., 2016).

The potential and application of data generated by genome-wide association studies, however, goes far beyond the identification of genetic variants associated with a disorder (Maier, Visscher, Robinson, & Wray, 2018). The most significant genome-wide loci associated with complex traits have small effect sizes and explain a very modest fraction of the predicted genetic variance. A major accomplishment of large genome-wide association studies of common traits is the empirical demonstration that the susceptibility to common disorders is polygenic, i.e., driven by many genetic variants of small effect.

Methods aggregating information from these common genetic variants have been developed and used in a variety of research settings (as described in the recent review by Maier and colleagues (Maier et al., 2018)). They have allowed the identification and quantification of substantial genetic overlaps among schizophrenia, bipolar disorder, and major depression (Purcell et al., 2009; Wray et al., 2018) and differences in the genetic architecture between subtypes of bipolar disorder (Allardyce et al., 2017; Charney et al., 2017). These methods hold promise for research into the biological underpinnings of postpartum psychosis. However, postpartum psychosis is a rare, nonunivocally defined, *orphan* disorder that poses unique challenges and offers unique opportunities (Florio et al., 2016). In the next section, we explore some of the problems and potentials of genetic biomarker discovery in postpartum psychosis.

Future directions

Phenotypic heterogeneity has been a major issue in genomic discovery of complex disorders, including bipolar disorder. Postpartum psychosis represents a relatively homogeneous subphenotype of bipolar disorder. In no other circumstances can a severe psychiatric disorder be ascertained within a specific 4-week period. The close temporal link with childbirth (Di Florio, Forty, et al., 2013; Heron, McGuinness, Blackmore, Craddock, & Jones, 2008; Heron et al., 2007; Munk-Olsen et al., 2006), the abrupt onset (Brockington et al., 1981; Dean & Kendell, 1981; Kendell, Chalmers, & Platz, 1987), the lower incidence (Kendell et al., 1987; Munk-Olsen et al., 2006), and the greater familiality (Jones & Craddock, 2002) suggest that postpartum psychosis represents a more homogeneous disease entity than bipolar disorder. It therefore offers a great opportunity to study a specific subtype of severe, pathological mood states, reducing the noise usually observed in studies involving complex, heterogenous disorders.

The need of large samples for genome-wide studies is probably the most significant limiting step in genetic biomarker discovery for postpartum psychosis. Affecting about 1 delivery in a 1000 (Kendell et al., 1987; Munk-Olsen et al., 2006), postpartum psychosis is about 100 times less common than postpartum depression (Costs of perinatal mental health problems, n.d.; Munk-Olsen et al., 2006). International consortia aggregating genetic data (What is the PGC?, n.d.), along with new methods of ascertainment, including electronic medical records (Castro et al., 2015) and smartphone applications dedicated to research (Guintivano et al., 2018), can help achieve the statistical power required for genome-wide investigations.

In summary, genetic biomarkers of postpartum psychosis have not been discovered yet. Postpartum psychosis represents a great opportunity for genomic research into the trigger of severe, pathological mood states. New technologies in the field of molecular genetics, large-scale phenotyping, and data sharing may boost research in this area.

References

Allardyce, J., Leonenko, G., Hamshere, M., Pardinas, A., Forty, L., Knott, S., … Escott-Price, V. (2017). Psychosis and the level of mood incongruence in bipolar disorder are related to genetic liability for Schizophrenia. *BioRxiv*, 160119 https://doi.org/10.1101/160119.

Bauer, A. E., Maegbaek, M. L., Liu, X., Wray, N. R., Sullivan, P. F., Miller, W. C., … Munk-Olsen, T. (2018). Familiality of psychiatric disorders and risk of postpartum psychiatric episodes: A population-based cohort study. *The American Journal of Psychiatry, 175*(8), 783–791. https://doi.org/10.1176/appi.ajp.2018.17111184.

Bipolar Disorder Working Group of the Psychiatric Genomics Consortium. (2019). Genome-wide association study identifies 30 loci associated with bipolar disorder. *Nature Genetics, 51*, 793–803.

Boyle, E. A., Li, Y. I., & Pritchard, J. K. (2017). An expanded view of complex traits: From polygenic to omnigenic. *Cell, 169*(7), 1177–1186. https://doi.org/10.1016/j.cell.2017.05.038.

Brockington, I. F., Cernik, K. F., Schofield, E. M., Downing, A. R., Francis, A. F., & Keelan, C. (1981). Puerperal psychosis. Phenomena and diagnosis. *Archives of General Psychiatry, 38*(7), 829–833.

Brockington, I. F., Martin, C., Brown, G. W., Goldberg, D., & Margison, F. (1990). Stress and puerperal psychosis. *The British Journal of Psychiatry: The Journal of Mental Science, 157*, 331–334.

Castro, V. M., Minnier, J., Murphy, S. N., Kohane, I., Churchill, S. E., Gainer, V., … International Cohort Collection for Bipolar Disorder Consortium. (2015). Validation of electronic health record phenotyping of bipolar disorder cases and controls. *The American Journal of Psychiatry, 172*(4), 363–372. https://doi.org/10.1176/appi.ajp.2014.14030423.

Charney, A. W., Ruderfer, D. M., Stahl, E. A., Moran, J. L., Chambert, K., Belliveau, R. A., … Sklar, P. (2017). Evidence for genetic heterogeneity between clinical subtypes of bipolar disorder. *Translational Psychiatry, 7*(1), e993. https://doi.org/10.1038/tp.2016.242.

Costs of perinatal mental health problems. (n.d.). Retrieved from September 10, 2017, Centre For Mental Health website https://www.centreformentalhealth.org.uk/costs-of-perinatal-mh-problems.

Coyle, N., Jones, I., Robertson, E., Lendon, C., & Craddock, N. (2000). Variation at the serotonin transporter gene influences susceptibility to bipolar affective puerperal psychosis. *Lancet (London, England), 356*(9240), 1490–1491. https://doi.org/10.1016/S0140-6736(00)02877-4.

Craddock, N., Brockington, I., Mant, R., Parfitt, F., McGuffin, P., & Owen, M. (1994). Bipolar affective puerperal psychosis associated with consanguinity. *The British Journal of Psychiatry: The Journal of Mental Science, 164*(3), 359–364.

Dean, C., & Kendell, R. E. (1981). The symptomatology of puerperal illnesses. *The British Journal of Psychiatry: The Journal of Mental Science, 139*, 128–133.

Dean, C., Williams, R. J., & Brockington, I. F. (1989). Is puerperal psychosis the same as bipolar manic-depressive disorder? A family study. *Psychological Medicine, 19*(3), 637–647.

Di Florio, A., Forty, L., Gordon-Smith, K., Heron, J., Jones, L., Craddock, N., & Jones, I. (2013). Perinatal episodes across the mood disorder spectrum. *JAMA Psychiatry, 70*(2), 168–175. https://doi.org/10.1001/jamapsychiatry.2013.279.

Di Florio, A., Jones, L., Forty, L., Gordon-Smith, K., Craddock, N., & Jones, I. (2015). Bipolar disorder, miscarriage, and termination. *Bipolar Disorders, 17*(1), 102–105. https://doi.org/10.1111/bdi.12217.

Di Florio, A., Smith, S., & Jones, I. (2013). Postpartum psychosis. *The Obstetrician and Gynaecologist, 15*(3), 145–150. https://doi.org/10.1111/tog.12041

Diagnostic and statistical manual of mental disorders. DSM Library. Retrieved from September 10, 2017 http://dsm.psychiatryonline.org.libproxy.lib.unc.edu/doi/book/10.1176/appi.books.9780890425596. (n.d.).

Dowlatshahi, D., & Paykel, E. S. (1990). Life events and social stress in puerperal psychoses: Absence of effect. *Psychological Medicine, 20*(3), 655–662.

Feng, J., Yan, J., Michaud, S., Craddock, N., Jones, I. R., Cook, E. H., … Sommer, S. S. (2001). Scanning of estrogen receptor alpha (ERalpha) and thyroid hormone receptor alpha (TRalpha) genes in patients with psychiatric diseases: Four missense mutations identified in ERalpha gene. *American Journal of Medical Genetics, 105*(4), 369–374.

Florio, A. D., Gordon-Smith, K., Forty, L., Kosorok, M. R., Fraser, C., Perry, A., … Jones, I. (2018). Stratification of the risk of bipolar disorder recurrences in pregnancy and postpartum. *The British Journal of Psychiatry, 213*(3), 542–547. https://doi.org/10.1192/bjp.2018.92.

Florio, A. D., Munk-Olsen, T., & Bergink, V. (2016). The birth of a psychiatric orphan disorder: Postpartum psychosis. *The Lancet Psychiatry, 3*(6), 502. https://doi.org/10.1016/S2215-0366(16)30037-2.

Guintivano, J., Krohn, H., Lewis, C., Byrne, E. M., Henders, A. K., Ploner, A., … Meltzer-Brody, S. (2018). PPD ACT: An app-based genetic study of postpartum depression. *Translational Psychiatry, 8*(1), 260. https://doi.org/10.1038/s41398-018-0305-5.

Heron, J., McGuinness, M., Blackmore, E. R., Craddock, N., & Jones, I. (2008). Early postpartum symptoms in puerperal psychosis. *BJOG: An International Journal of Obstetrics and Gynaecology, 115*(3), 348–353. https://doi.org/10.1111/j.1471-0528.2007.01563.x.

Heron, J., Robertson Blackmore, E., McGuinness, M., Craddock, N., & Jones, I. (2007). No "latent period" in the onset of bipolar affective puerperal psychosis. *Archives of Women's Mental Health, 10*(2), 79–81. https://doi.org/10.1007/s00737-007-0174-z.

Hou, L., Heilbronner, U., Degenhardt, F., Adli, M., Akiyama, K., Akula, N., … Schulze, T. G. (2016). Genetic variants associated with response to lithium treatment in bipolar disorder: A genome-wide association study. *The Lancet, 387*(10023), 1085–1093. https://doi.org/10.1016/S0140-6736(16)00143-4.

Huhn, A., & Drenk, K. (1973). [Clinical classification and prognosis of postpartum-psychoses]. *Fortschritte der Neurologie, Psychiatrie und Ihrer Grenzgebiete, 41*(7), 363–377.

ICD-10 Version:2010. (n.d.). Retrieved from August 30, 2018 http://apps.who.int/classifications/icd10/browse/2010/en.

International Schizophrenia Consortium, Purcell, S. M., Wray, N. R., Stone, J. L., Visscher, P. M., O'Donovan, M. C., … Sklar, P. (2009). Common polygenic variation contributes to risk of schizophrenia and bipolar disorder. *Nature, 460*(7256), 748–752. https://doi.org/10.1038/nature08185.

Jones, I., & Craddock, N. (2001). Familiality of the puerperal trigger in bipolar disorder: Results of a family study. *The American Journal of Psychiatry, 158*(6), 913–917. https://doi.org/10.1176/appi.ajp.158.6.913.

Jones, I., & Craddock, N. (2002). Do puerperal psychotic episodes identify a more familial subtype of bipolar disorder? Results of a family history study. *Psychiatric Genetics, 12*(3), 177–180.

Jones, I., Hamshere, M., Nangle, J.-M., Bennett, P., Green, E., Heron, J., … Craddock, N. (2007). Bipolar affective puerperal psychosis: Genome-wide significant evidence for linkage to chromosome 16. *The American Journal of Psychiatry, 164*(7), 1099–1104. https://doi.org/10.1176/ajp.2007.164.7.1099.

Jones, I., Middle, F., McCandless, F., Coyle, N., Robertson, E., Brockington, I., … Craddock, N. (2000). Molecular genetic studies of bipolar disorder and puerperal psychosis at two polymorphisms in the estrogen receptor alpha gene (ESR 1). *American Journal of Medical Genetics, 96*(6), 850–853.

Kadrmas, A., Winokur, G., & Crowe, R. (1979). Postpartum mania. *The British Journal of Psychiatry: The Journal of Mental Science, 135*, 551–554.

Kendell, R. E., Chalmers, J. C., & Platz, C. (1987). Epidemiology of puerperal psychoses. *The British Journal of Psychiatry: The Journal of Mental Science, 150*, 662–673.

Maier, R. M., Visscher, P. M., Robinson, M. R., & Wray, N. R. (2018). Embracing polygenicity: A review of methods and tools for psychiatric genetics research. *Psychological Medicine, 48*(7), 1055–1067. https://doi.org/10.1017/S0033291717002318.

Marks, M. N., Wieck, A., Checkley, S. A., & Kumar, R. (1991). Life stress and post-partum psychosis: A preliminary report. *British Journal of Psychiatry — Supplement*, (10), 45—49.

Middle, F., Jones, I., Robertson, E., Morey, J., Lendon, C., & Craddock, N. (2003). Variation in the coding sequence and flanking splice junctions of the estrogen receptor alpha (ERalpha) gene does not play an important role in genetic susceptibility to bipolar disorder or bipolar affective puerperal psychosis. *American Journal of Medical Genetics Part B, Neuropsychiatric Genetics: The Official Publication of the International Society of Psychiatric Genetics, 118B*(1), 72—75. https://doi.org/10.1002/ajmg.b.10021.

Munk-Olsen, T., Laursen, T. M., Meltzer-Brody, S., Mortensen, P. B., & Jones, I. (2012). Psychiatric disorders with postpartum onset: Possible early manifestations of bipolar affective disorders. *Archives of General Psychiatry, 69*(4), 428—434. https://doi.org/10.1001/archgenpsychiatry.2011.157.

Munk-Olsen, T., Laursen, T. M., Mendelson, T., Pedersen, C. B., Mors, O., & Mortensen, P. B. (2009). Risks and predictors of readmission for a mental disorder during the postpartum period. *Archives of General Psychiatry, 66*(2), 189—195. https://doi.org/10.1001/archgenpsychiatry.2008.528.

Munk-Olsen, T., Laursen, T. M., Pedersen, C. B., Mors, O., & Mortensen, P. B. (2006). New parents and mental disorders: A population-based register study. *Journal of the American Medical Association, 296*(21), 2582—2589. https://doi.org/10.1001/jama.296.21.2582.

Nakigudde, J., Ehnvall, A., Mirembe, F., Musisi, S., & Airaksinen, E. (2013). An exploratory study on the feasibility and appropriateness of family psychoeducation for postpartum women with psychosis in Uganda. *BMC Psychiatry, 13*, 131. https://doi.org/10.1186/1471-244X-13-131.

Osterman, E. (1963). Post partum psychopathologic states. *Acta Psychiatrica Scandinavica, 39*(Suppl. 169), 190—192.

Perry, A., Gordon-Smith, K., Di Florio, A., Forty, L., Craddock, N., Jones, L., & Jones, I. (2016). Adverse childhood life events and postpartum psychosis in bipolar disorder. *Journal of Affective Disorders, 205*, 69—72. https://doi.org/10.1016/j.jad.2016.06.061.

Platz, C., & Kendell, R. E. (1988). A matched-control follow-up and family study of "puerperal psychoses". *The British Journal of Psychiatry: The Journal of Mental Science, 153*, 90—94.

Posmontier, B., & Fisher, K. M. (2014). A narratology of postpartum psychosis in an Orthodox Jewish woman. *Perspectives in Psychiatric Care, 50*(3), 167—177. https://doi.org/10.1111/ppc.12037.

Protheroe, C. (1969). Puerperal psychoses: A long term study 1927—1961. *The British Journal of Psychiatry: The Journal of Mental Science, 115*(518), 9—30.

Purcell, S. M., Wray, N. R., Stone, J. L., Visscher, P. M., O'Donovan, M. C., Sullivan, P. F., … Moran, J. L. (2009). Common polygenic variation contributes to risk of schizophrenia and bipolar disorder. *Nature, 460*(7256), 748—752. https://doi.org/10.1038/nature08185.

Ripke, S., Neale, B. M., Corvin, A., Walters, J. T., Farh, K.-H., Holmans, P. A., … O'Donovan, M. C. (2014). Biological insights from 108 schizophrenia-associated genetic loci. *Nature, 511*(7510), 421—427. https://doi.org/10.1038/nature13595.

Ruderfer, D. M., Fanous, A. H., Ripke, S., McQuillin, A., Amdur, R. L., Schizophrenia Working Group of the Psychiatric Genomics Consortium, … Kendler, K. S. (2014). Polygenic dissection of diagnosis and clinical dimensions of bipolar disorder and schizophrenia. *Molecular Psychiatry, 19*(9), 1017—1024. https://doi.org/10.1038/mp.2013.138.

Schöpf, J., Bryois, C., Jonquiére, M., & Scharfetter, C. (1985). A family hereditary study of post-partum "psychoses." *European Archives of Psychiatry and Neurological Sciences, 235*(3), 164—170.

Schöpf, J., & Rust, B. (1994). Follow-up and family study of postpartum psychoses. Part I: Overview. *European Archives of Psychiatry and Clinical Neuroscience, 244*(2), 101–111.

Seager, C. P. (1960). A controlled study of post-partum mental illness. *The Journal of Mental Science, 106*, 214–230.

Starnawska, A., Demontis, D., Pen, A., Hedemand, A., Nielsen, A. L., Staunstrup, N. H., … Nyegaard, M. (2016). CACNA1C hypermethylation is associated with bipolar disorder. *Translational Psychiatry, 6*(6), e831. https://doi.org/10.1038/tp.2016.99.

Thippeswamy, H., Dahale, A., Desai, G., & Chandra, P. S. (2015). What is in a name? Causative explanatory models of postpartum psychosis among patients and caregivers in India. *The International Journal of Social Psychiatry, 61*(8), 818–823. https://doi.org/10.1177/0020764015597016.

Thuwe, I. (1974). Genetic factors in puerperal psychosis. *The British Journal of Psychiatry: The Journal of Mental Science, 125*(0), 378–385.

VanderKruik, R., Barreix, M., Chou, D., Allen, T., Say, L., Cohen, L. S., & Maternal Morbidity Working Group. (2017). The global prevalence of postpartum psychosis: A systematic review. *BMC Psychiatry, 17*(1), 272. https://doi.org/10.1186/s12888-017-1427-7.

Whalley, L. J., Roberts, D. F., Wentzel, J., & Wright, A. F. (1982). Genetic factors in puerperal affective psychoses. *Acta Psychiatrica Scandinavica, 65*(3), 180–193.

What is the PGC? — Psychiatric Genomics Consortium [Page]. (n.d.). Retrieved from December 7, 2018 http://www.med.unc.edu/pgc.

Wray, N. R., Lee, S. H., Mehta, D., Vinkhuyzen, A. A. E., Dudbridge, F., & Middeldorp, C. M. (2014). Research review: Polygenic methods and their application to psychiatric traits. *The Journal of Child Psychology and Psychiatry and Allied Disciplines, 55*(10), 1068–1087. https://doi.org/10.1111/jcpp.12295.

Wray, N. R., Ripke, S., Mattheisen, M., Trzaskowski, M., Byrne, E. M., Abdellaoui, A., … Major Depressive Disorder Working Group of the Psychiatric Genomics Consortium. (2018). Genome-wide association analyses identify 44 risk variants and refine the genetic architecture of major depression. *Nature Genetics, 50*(5), 668–681. https://doi.org/10.1038/s41588-018-0090-3.

CHAPTER 12

Hormonal and immunological factors in postpartum psychosis

Jamie Maguire
Tufts University School of Medicine, Department of Neuroscience, Boston, MA, United States of America

Introduction

Postpartum psychosis is clinically defined as the presence of psychotic symptoms (including delusions, hallucinations, disorganized thinking, paranoia, disorganized or abnormal motor behavior, agitation), usually accompanied by affective symptoms (manic, depressive, or mixed). The Diagnostic and Statistical Manual Version 5 (DSM-5) does not recognize postpartum psychosis as a distinct disorder, but rather uses the specifier of "with peripartum onset" added to a diagnosis of brief psychotic disorder if the episode begins during pregnancy or within 4 weeks postpartum. The symptoms of postpartum psychosis are severe and have a rapid onset postpartum, resulting in a 70% increase in the risk for suicide (for review see Sit, Rothschild, & Wisner, 2006) and a 4% risk of infanticide (for review see Spinelli, 2009).

There is agreement in the literature regarding the incidence of postpartum psychosis, which is reported to occur in approximately 1–5 out of every 1000 deliveries in the general population (VanderKruik et al., 2017). Risk factors for developing postpartum psychosis include a history of psychotic or bipolar episodes, family history of postpartum psychosis or bipolar disorder, stressful life experiences, lack of support, sleep deprivation, prenatal infant mortality, and discontinuation of mood stabilizers (Table 12.1) (Kendell, Chalmers, & Platz, 1987; McNeil, 1987; Nonacs & Cohen, 1998). In addition to these predisposing factors, both biological and psychosocial aspects of the peripartum period are thought to confer vulnerability to the development of postpartum psychosis. For instance, there is an increased incidence of psychosis during the peripartum period compared to other periods in women's lives (Kendell et al., 1987; Spinelli, 2009). Therefore, changes that occur during the postpartum period have become a focus for investigation into potential triggers and/or causes of postpartum psychosis. As such, the hormonal changes and

Biomarkers of Postpartum Psychiatric Disorders
ISBN 978-0-12-815508-0
https://doi.org/10.1016/B978-0-12-815508-0.00012-6

Table 12.1 Risk Factors for postpartum psychosis.

History of psychotic episode
History of bipolar episode
Family history of postpartum psychosis
Family history of bipolar disorder
Previous stressful life experiences
Lack of partner/family support
Sleep deprivation
Prenatal infant mortality
Discontinuation of mood-stabilizing drugs

immunological changes that occur during the peripartum period have received a great deal of attention and will be the focus of this chapter.

Hormonal changes associated with postpartum psychosis

The peripartum period is a time of dramatic changes in hormone levels (for review see Tal, Taylor, Burney, Mooney, & Giudice, 2000), and these changes are thought to play a role in the vulnerability to mood disorders during this period. A coordinated sequence of events, including well-orchestrated neuroendocrine changes, is required to establish and maintain a successful pregnancy. The characteristic neuroendocrine changes that accompany pregnancy will be discussed in this chapter, with a particular focus on progesterone, estrogen, oxytocin, prolactin, and cortisol, and are summarized in Fig. 1. Many of the endocrine changes accompanying pregnancy occur at the boundary between the mother and fetus, the feto-placental unit (FPU), which is a major site of steroid hormone production and secretion. During early pregnancy, progesterone is produced largely by the corpus luteum, an endocrine gland consisting of a mass of cells formed from a follicle in the ovary, rising to supraphysiological levels (approximately a 10-fold increase throughout pregnancy) to support pregnancy following conception. During late pregnancy, progesterone is largely produced by the placenta, and it functions to suppress the maternal immunologic response and to prevent preterm labor. The placenta is also the major source of estrogen production throughout pregnancy; levels of estrogen increase to 100-fold the levels in nonpregnant women. Prolactin increases during late pregnancy, reaching levels 10–20 times baseline, and

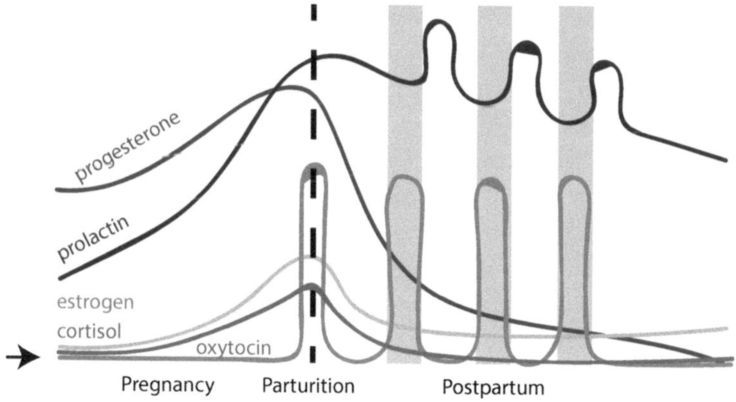

Figure 1 Fluctuations in reproductive/lactogenic hormones throughout the peripartum period.

remains elevated throughout lactation. Oxytocin levels also increase during late pregnancy and peak during delivery; however, it is interesting to note that the changes in the levels of oxytocin appear to be more variable compared to those in other steroid hormones during the peripartum period. Cortisol is also synthesized by the placenta, and this placental cortisol serves to suppress the maternal immune response that is thought to confer immunologic privilege to the fetus. Despite the increased levels of secreted corticotropin-releasing hormone (CRH), adrenocorticotropic hormone (ACTH), and cortisol, pregnant women do not exhibit hypercortisolemia. In fact, the stimulated or stress-induced activation of the hypothalamic–pituitary–adrenal (HPA) axis is suppressed during pregnancy and lactation due to negative feedback regulation of the HPA axis (Brunton, Russell, & Douglas, 2008; Brunton & Russell, 2008, 2011; De Weerth & Buitelaar, 2005; Schulte, Weisner, & Allolio, 1990).

Careful measurement of steroid hormone levels throughout pregnancy and the postpartum period demonstrates increases in serum estradiol, 17α-hydroxyprogesterone, cortisol, and 11-deoxycortisol throughout pregnancy, as well as a plateau in androstenedione levels; all of these hormones then decrease dramatically postpartum (Soldin et al., 2005). In contrast, the endogenous steroid hormone dehydroepiandrosterone (DHEA) and its sulfated form dehydroepiandrosterone sulfate (DHEAS) remain relatively constant throughout the peripartum period (Soldin et al., 2005). The abrupt decrease in steroid hormone levels postpartum is thought to precipitate mood disorders and may even directly contribute to the underlying neurobiology of postpartum disorders, such as postpartum depression and postpartum psychosis. However, a critical study demonstrated that withdrawal of hormones alone is not sufficient to induce depression symptoms in all women. Rather, hormone withdrawal only induces depression symptoms in patients with a history of postpartum depression (Bloch et al., 2000), suggesting that it is not withdrawal of hormones per se but that the sensitivity to hormone withdrawal may be altered in patients with postpartum depression. Although a similar study has not been conducted specifically for postpartum psychosis, it is likely that there is also a predisposition to the impact of hormone withdrawal associated with postpartum psychosis. The following section will review steroid hormone signaling implicated in postpartum disorders as demonstrated in Fig. 2.

Estrogen

Estrogen signaling is mediated by several different receptor subtypes, including estrogen receptors, ERα and ERβ, as well as the G-protein–coupled estrogen receptor 1 (GPER1). Estrogen signaling involves alterations in gene transcription, while signaling through GPER1 can also exert nongenomic actions via activation of intracellular signaling cascades. As mentioned above, estrogen levels increase dramatically

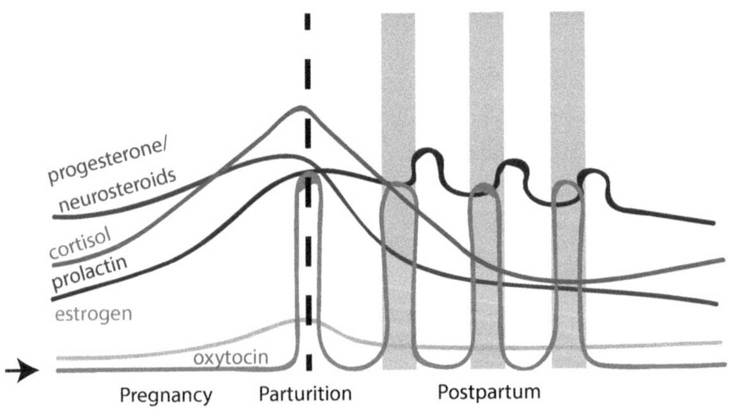

Figure 2 Alterations in reproductive/lactogenic hormones implicated in postpartum psychosis.

throughout pregnancy, reaching levels 100 times their baseline levels, and then decreasing precipitously after parturition.

Although the potential role of estrogen signaling in postpartum psychiatric disorders is not mechanistically clear, there is some evidence of disruptions in estrogen signaling in both postpartum depression and postpartum psychosis. Perhaps the strongest evidence is the therapeutic effect of estrogen treatment on postpartum mood disorders (Ahokas, Kaukoranta, Wahlbeck, & Aito, 2001; Gregoire, Kumar, Everitt, & Studd, 1996; Lawrie, Herxheimer, & Dalton, 2000; Sichel, Cohen, Robertson, Ruttenberg, & Rosenbaum, 1995; Walsh, Seidman, Sysko, & Gould, 2002) (for review see Moses-Kolko, Berga, Kalro, Sit, & Wisner, 2009).

The extensive literature on schizophrenia may help to inform our understanding of the role of estrogen in postpartum psychosis. The concept that ovarian hormones may influence the underlying neurobiology of schizophrenia comes from the evidence of sex differences in the incidence of schizophrenia, with the male:female incidence at approximately 1.4:1 (for review see Abel, Drake, & Goldstein, 2010), which has been attributed to differences in gonadal hormones (for review see Gogos et al., 2015). The estrogen hypothesis has been the most extensively explored and suggests that estrogen plays a protective role in schizophrenia, blunting the severity of the symptoms (for review see Gogos et al., 2015). Mechanistically, this has been suggested to involve the ability of estrogen to affect mood, cognition, and behavior (Fink, Sumner, Rosie, Grace, & Quinn, 1996). Consistent with a role for estrogen in schizophrenia, estrogen levels have been found to be lower in women with schizophrenia (Bergemann et al., 2005; Huber et al., 2001). Furthermore, women with schizophrenia exhibit menstrual cycle—linked changes in symptoms, with an increase in symptoms immediately preceding or during menstruation when estrogen levels are low (Dalton, 1959) and symptoms subside during pregnancy (Chang & Renshaw, 1986) and worsen during the postpartum period (Kendell et al., 1987) following the precipitous decline in estrogen and progesterone. Treatment with estrogen has shown promising results in clinical trials, particularly in treating the positive symptoms of schizophrenia (Akhondzadeh et al., 2003; Kulkarni et al., 2001, 2014). These data suggest that ovarian hormones may play a role in the psychotic symptoms of schizophrenia as well as in postpartum psychosis.

Patients with postpartum onset of psychosis exhibit baseline serum estradiol levels that are lower than the threshold for gonadal failure (Ahokas & Aito, 1999; Ahokas, Aito, & Rimon, 2000). Treatment with 17-β estradiol for

6 weeks, which increased serum estradiol levels, improved symptoms in patients diagnosed with postpartum psychosis (Ahokas & Aito, 1999; Ahokas et al., 2000). It is likely that there are diverse etiologies of patients suffering from postpartum psychosis and a subset of patients presenting with low serum estradiol levels may benefit from treatment with 17-β estradiol.

Progesterone and allopregnanolone

The actions of progesterone are exerted both through classical signaling via nuclear progesterone receptors (PRs), which activate genes containing progesterone response elements (PREs), and through diverse nonclassical signaling mechanisms involving membrane progesterone receptors (mPRs) and activation of intracellular signaling cascades (for review see Garg, Ng, Baig, Driggers, & Segars, 2017).

The relationship between progesterone and postpartum mood disorders is thought to involve, at least in part, the effects of progesterone on $GABA_A$ receptors. Neuroactive metabolites of progesterone, in particular allopregnanolone, can modulate $GABA_A$ receptors. Allopregnanolone acts as a positive allosteric modulator at $GABA_A$ receptors, enhancing the inhibitory effects of GABA, and therefore has been the focus of investigation into the potential role for allopregnanolone in contributing to numerous neurological and neuropsychiatric disorders (MacKenzie & Maguire, 2013), including postpartum depression (for a review see Schüle, Nothdurfter, & Rupprecht, 2014). It is well established that allopregnanolone is capable of exerting anxiolytic and antidepressant effects (Schüle et al., 2014), supporting a potential role for allopregnanolone in mood disorders, particularly related to the peripartum period. The dramatic increase in progesterone levels throughout pregnancy followed by the abrupt decrease following parturition has led to the hypothesis that deficits in progesterone and/or allopregnanolone signaling might precipitate postpartum disorders. However, there have been inconsistent findings regarding the levels of progesterone or the neuroactive metabolite, allopregnanolone, in patients with postpartum mood disorders. Recent evidence suggests that lower levels of allopregnanolone during pregnancy predict the development of postpartum depression (Osborne et al., 2017) and perhaps the most compelling evidence comes from the therapeutic effects of treatment with a synthetic formulation of allopregnanolone, brexanolone, in patients suffering from postpartum depression (S. Kanes et al., 2017; S. J. Kanes et al., 2017; Meltzer-Brody et al., 2018). However, the relationship to postpartum psychosis is less clear given the limited investigations related to this topic.

As discussed above in relation to estrogen, gonadal hormones have been suggested to influence the underlying neurobiology of schizophrenia due to the evidence of sex differences and menstrual cycle–linked changes in symptom presentation (for review see Abel et al., 2010). The same premise holds for changes in progesterone. In fact, it has been pointed out that the importance of progesterone in schizophrenia has been neglected (Sun, Walker, Dean, van den Buuse, & Gogos, 2016). The association between progesterone levels and schizophrenia has produced mixed results (for review see Sun et al., 2016). In addition to a direct role for progesterone on schizophrenia, neuroactive derivatives of progesterone, including pregnanolone and allopregnanolone, have been shown to be altered in schizophrenia (Marx et al., 2005) and are proposed to be promising treatments for schizophrenia (Marx et al., 2011). These results suggest that progesterone and/or neurosteroids may play a role in psychosis; given the fluctuations of these hormones during the peripartum period and evidence of their contributions in both affective and psychotic disorders, there is a plausible role in postpartum psychosis as well.

Interestingly, there is an older case study in which progesterone treatment was used to treat postpartum psychosis (Schmidt, 1943). The patient presented with postpartum psychosis, which upon return of menstruation followed a pattern of cyclic recurrence with psychosis in the latter half of the menstrual cycle. Assuming that this pattern was attributed to a hormonal imbalance, the patient was treated with progesterone, which improved the postpartum psychosis (Schmidt, 1943). Collectively, these studies implicate altered progesterone and/or its neuroactive metabolite, allopregnanolone, in contributing to the underlying neurobiology of postpartum disorders, including postpartum depression and postpartum psychosis.

Oxytocin

Oxytocin signaling is largely mediated by its main target, the oxytocin receptor (OTR), a G-protein–coupled receptor that exerts a wide range of actions, including induction of proinflammatory cytokines, osteoclastogenesis, nociception, and induction of anxiolytic and cardioprotective effects (for review see Chatterjee et al., 2016). In addition, oxytocin can also bind to and activate vasopressin receptors (V1a, V1b, and V2), further increasing the diversity in oxytocin signaling.

Oxytocin levels are increased during the peripartum period and play a critical role in regulating mood and mother–infant interactions (Bell, Erickson, & Carter, 2014). Reductions in oxytocin levels have been suggested to predict the development of postpartum depression (Skrundz, Bolten, Nast, Hellhammer, & Meinlschmidt, 2011) and are correlated with depressive symptoms (Stuebe, Grewen, & Meltzer-Brody, 2013). However, intranasal oxytocin failed to improve maternal caregiving behaviors (Mah, Van Ijzendoorn, Out, Smith, & Bakermans-Kranenburg, 2017) and may even increase the incidence of postpartum depression (Kroll-Desrosiers et al., 2017). These data highlight the complexity of hormonal factors in the etiology of postpartum depression.

Although there does not appear to be any direct evidence linking oxytocin and postpartum psychosis, oxytocin has been suggested to have potential for the treatment of schizophrenia. Emerging evidence suggests that oxytocin levels may correlate with symptoms of schizophrenia and oxytocin treatment may be beneficial for the treatment of both the positive and negative symptoms of schizophrenia (for review see Rich & Caldwell, 2015 and Shilling & Feifel, 2016). Intranasal oxytocin decreased both the positive and negative symptoms of schizophrenia (for review see Rich & Caldwell, 2015 and Shilling & Feifel, 2016). Although the mechanisms through which oxytocin influences the symptoms of schizophrenia are unclear, the well-established role of oxytocin in regulating emotion, social interaction, stress, and attachment (Bell et al., 2014) is thought to be involved in improving the negative symptoms of schizophrenia; whereas the proposed role of oxytocin as a cognitive enhancer and potential role in plasticity is thought to mediate the effects on the positive symptoms of schizophrenia. As such, oxytocin may also have potential therapeutic benefits for postpartum psychosis. However, to our knowledge this has not been directly investigated to-date.

Prolactin

The prolactin signaling pathway is initiated by the binding of prolactin to its prolactin receptor (PRLR), which is a member of the class I cytokine receptor superfamily. Activation of the PRLR primarily involves activation of the JAK/STAT pathway, although prolactin also activates the MAP kinase pathway, the PI3-Kinase pathway, and the RAC pathway (for review see Radhakrishnan et al., 2012).

Prolactin has been implicated in postpartum disorders, given that the levels of prolactin are increased during the peripartum period and the fact that prolactin plays a role in lactogenesis and maternal behaviors (for review see Stuebe, Grewen, Pedersen, Propper, & Meltzer-Brody, 2012). Women with postpartum depression, even those that are breastfeeding, exhibit a decrease in prolactin levels (Groer & Morgan, 2007; Harris et al., 1989), and decreased prolactin levels correlate with depression scores in women with postpartum depression (Abou-Saleh, Ghubash, Karim, Krymski, & Bhai, 1998).

Although there is no direct evidence for a role for prolactin in postpartum psychosis, similar to oxytocin, prolactin has been implicated in schizophrenia and other psychoses. There is a negative correlation between plasma prolactin levels and the positive symptoms of schizophrenia (for review see Rajkumar, 2014). Furthermore, prolactin secretion is increased by antipsychotic drugs and is thought to play a role in treatment response (for review see Rajkumar, 2014). Thus, although there are no direct reports of a role for prolactin in postpartum psychosis, these studies suggest that altered prolactin levels can influence psychotic symptoms, and given the fluctuations in prolactin levels during the peripartum period it is feasible that prolactin could affect psychosis with postpartum onset.

Stress hormones: CRH, ACTH, and cortisol

In addition to ovarian and lactogenic hormones, stress hormones are altered during the peripartum period and have also been implicated in the development of postpartum disorders, particularly postpartum depression. This is based, in part, on the evidence of neuroendocrine disruptions and hypercortisolemia associated with major depressive disorder (Pariante & Lightman, 2008). The most convincing data surround the role of HPA axis dysfunction, which governs the body's physiological response to stress, in postpartum depression (for review see Bloch et al., 2003bib_-Bloch_et_al_2003; Brummelte & Galea, 2010; Glynn et al., 2013bib_-Glynn_et_al_2013; Magiakou et al., 1996). Stress and previous adverse life events are major risk factors for developing postpartum mood disorders (Guintivano et al., 2018), and there is evidence for altered stress reactivity in women with postpartum mood disorders (Bloch et al., 2003, 2005). However, the mechanisms whereby HPA axis disruption would impact postpartum psychiatric disorders are unclear.

Altered glucocorticoid signaling has also been implicated in psychosis, including postpartum psychosis. Patients with schizophrenia and bipolar disorder exhibit an increased waking cortisol response (Girshkin, Matheson, Shepherd, & Green, 2014) as do individuals at a high risk for schizophrenia (Walker et al., 2013) or bipolar disorder (Ostiguy, Ellenbogen, Walker, Walker, & Hodgins, 2011). Patients at risk for developing psychosis also exhibit HPA axis hyperactivity (Garner et al., 2005; Mittal, Dhruv, Tessner, Walder, & Walker, 2007) and patients exhibiting psychotic symptoms have altered HPA axis responsiveness (for review see Corcoran et al., 2003; Pariante, 2008, 2009). Perhaps the most convincing evidence for a role for glucocorticoids in psychosis is the fact that a side effect of excessive glucocorticoid exposure, particularly related to steroid abuse, is "steroid psychosis" (for review see Dubovsky, Arvikar, Stern, & Axelrod, 2012). Thus, altered glucocorticoid signaling is capable of inducing psychotic symptoms, and although this has not been investigated in postpartum psychosis specifically, it is possible that dysregulation of the HPA axis during the peripartum period may increase vulnerability to psychosis.

It has been suggested that glucocorticoid resistance plays a role in postpartum disorders, including postpartum psychosis (for review see Pariante, 2009). To understand glucocorticoid resistance, one must first understand glucocorticoid signaling. Glucocorticoids, including cortisol, signal through mineralocorticoid receptors (MRs) and glucocorticoid receptors (GRs). MRs and GRs are members of the steroid receptor family of ligand-activated transcription factors and exert their effects primarily through altering gene transcription of effector proteins. In addition to their classic genomic effects, glucocorticoids can also exert rapid, nongenomic effects via several intracellular signaling pathways involving phosphoinositide 3-kinase (PI3K), protein kinase B (AKT), and mitogen-activated protein kinases (MAPKs) (for review see Oakley & Cidlowski, 2013).

Altered negative feedback of glucocorticoids, consistent with glucocorticoid resistance, is supported by the evidence of increased pituitary volume in patients with psychosis (Pariante, 2008) and increased inflammation (Zhang, Zhou, Cao, Wu, & Shen, 2005). However, there are very few studies investigating changes in glucocorticoid receptors associated with psychosis, although a few studies have demonstrated a reduction in GRs in patients with schizophrenia (Perlman, Webster, Kleinman, & Weickert, 2004; Yehuda, Boisoneau, Mason, & Giller, 1993). Interestingly, glucocorticoid resistance is also associated with increased inflammation in patients with psychiatric disorders (Raison, Capuron, & Miller, 2006). These data suggest that altered stress hormones may impact postpartum psychosis through their well-established role in regulating the immune system. The role of immunological changes associated with postpartum psychosis will be discussed in the subsequent section.

Thyroid hormones

Thyroid hormones (THs), including thyroxine (T3) and triiodothyronine (T4), are secreted by the thyroid gland and classically signal through TH receptors (THRs), which are encoded by two genes THRA and THRB which give rise to numerous receptor isoforms. THs acting on these receptors exert a range of effects classically through regulation of gene expression. In addition, THs also exert nongenomic effects, which can activate intracellular signaling cascades inducing numerous effects independent of gene transcription (for review see Harvey & Williams, 2002). Due to these diverse signaling mechanisms, thyroid hormones have been shown to influence a wide range of functions, including metabolism and homeostatic functions. Furthermore, dysregulation of thyroid hormones has been linked to anxiety, depression, and menstrual cycle dysregulation (Bathla, Singh, & Relan, 2016) with potential relevance to postpartum psychosis.

Postpartum thyroiditis (PPT) is a common postpartum condition, occurring in 2%–9% of postpartum women. Interestingly, a case study described a patient with simultaneously occurring PPT and postpartum psychosis, which suggested that there may be a shared underlying pathology (Bokhari, Bhatara, Bandettini, & Michael McMillin, 1998). In fact, postpartum psychosis has also been linked to primary hypoparathyroidism (for review see Bergink, Rasgon, & Wisner, 2016). Thyroid hormone dysfunction has been further implicated specifically in postpartum psychosis largely based on the increased incidence of autoimmune thyroid disease (AITD) in patients suffering with postpartum psychosis, suggesting that AITD is a potential etiological factor for the development of postpartum psychosis (Bergink et al., 2011, 2018). Collectively, these findings suggest that thyroid hormone dysfunction may be contributing to postpartum psychosis.

Consistent with a potential role in psychosis, thyroid hormones have also been implicated in schizophrenia (for review see Santos et al., 2012). Although the mechanisms whereby thyroid hormones influence schizophrenia are not well understood, it has been proposed to involve the ability of thyroid hormones to modulate numerous pathophysiological mechanisms, including neurotransmission, myelination, and inflammation (for review see Santos et al., 2012). Together, these studies implicate thyroid hormones in psychosis, including a role in both schizophrenia and postpartum psychosis.

Immunological changes associated with postpartum psychosis

There are limited studies investigating immunological changes associated with postpartum psychosis; there are more studies describing immune changes during the peripartum period associated with postpartum depression or in psychotic symptoms in patients with schizophrenia. Therefore, this section will review normal immune changes throughout the peripartum period and will briefly touch on changes implicated in depression and postpartum depression (see Chapter 1). Furthermore, this section will highlight immunological changes implicated in psychosis in association with schizophrenia with potential relevance to postpartum psychosis.

There are changes in the inflammatory state throughout the peripartum period, including an immunosuppressive state to allow fetal allograft implantation and characteristic immune changes including proinflammatory as well as antiinflammatory states (for review see Luppi, 2003; Mor & Cardenas, 2010; Mor et al., 2011). The proinflammatory state involves changes in chemokines, cytokines, and growth factors, such as an increase in interleukin-4 (IL-4), IL-5, IL-10, and transforming growth factor beta (TGF-β) and a decrease in interferon gamma (INF-γ), tumor necrosis factor-α (TNF-α), and IL-2. Whereas, the antiinflammatory state is orchestrated by placental changes altering the function of maternal immune cells. Throughout pregnancy, there are characteristic changes in immune cells with an increase in the number of regulatory T cells, natural killer cells, macrophages, and granulocytes. Dysregulation in these normal inflammatory changes may contribute to postpartum psychiatric illness.

How could these immune changes be related to neuropsychiatric illness? Cytokines are known to influence neurotransmission and neuroendocrine function (Miller, Haroon, Raison, & Felger, 2013), which can influence depression as summarized in "Hormonal changes associated with postpartum psychosis" section. Exposure to high levels of proinflammatory cytokines has been shown to induce changes in mood and other symptoms associated with depression, including fatigue, anxiety, sleep disturbances, and anhedonia (Capuron & Miller, 2004; Pollak & Yirmiya, 2002). A good example is the evidence that cytokines acting in the brain can induce "sickness behavior," a set of behavioral changes associated with infection, which can then develop into depression (Dantzer, O'Connor, Freund, Johnson, & Kelley, 2008).

Considerable evidence supports the role of immunological change in depression outside of the peripartum. Subsets of patients with major depressive disorder exhibit alteration in the peripheral immune system, including impairments in cellular immunity and increased levels of proinflammatory cytokines (for review see Felger & Lotrich, 2013; Zunszain et al., 2013), including IL-1, IL-6, and TNF-α (Dowlati et al., 2010; Howren, Lamkin, & Suls, 2009; Levine et al., 1999; Gimeno et al., 2009).

Given this association, it is plausible that alterations in proinflammatory cytokines during the peripartum period may contribute to the increased vulnerability to mood disorders during the postpartum period. In fact, postpartum depression has been suggested to be an immune disorder (Gleicher, 2007; Licinio, 2009), with levels of IL-6, IL-1β, and TNF-α positively correlated with depressive symptoms in antenatal and postpartum women (Boufidou et al., 2009; Cassidy-Bushrow, Peters, Johnson, & Templin, 2012; Corwin & Pajer, 2008). Interestingly, levels of IL-6 and TNF-α are also associated with previous adverse life events (Blackmore et al., 2011), a known risk factor for postpartum depression. T cells are also negatively correlated with depression in the postpartum period (Hucklebridge et al., 1994), similar to the relationship found in major depressive disorder (Grosse et al., 2016). These data provide evidence of a relationship between immunological changes and postpartum depression.

Inflammation has also been implicated in psychosis. For example, an increase in proinflammatory cytokines has been demonstrated in patients with schizophrenia (for review see Müller et al., 2015). A vulnerability—stress—inflammation model of schizophrenia has been proposed (for review see Müller et al., 2015), in which stress triggers psychotic episodes involving an underlying predisposition associated with inflammation. The relationship between inflammation and psychosis suggests that the immune changes which occur throughout the peripartum period may directly contribute or confer vulnerability to postpartum psychosis. Microglia have also been implicated in psychotic disorders, including schizophrenia. This hypothesis holds that microglia hyperactivation contributes to schizophrenia (Tay et al., 2017). Part of this theory suggests that early infection results in microglia hyperactivation into adulthood which predisposes individuals to psychosis, consistent with previous adverse life events as major risk factors for developing postpartum depression.

There is accumulating evidence for a role for immune system dysfunction and inflammation in postpartum psychosis (for review see Bergink, Gibney, & Drexhage, 2014). The high incidence of autoimmune conditions and chronic inflammation associated with postpartum psychosis implicates the immune system in the underlying neurobiology of postpartum psychosis (for review see Bergink et al., 2014). One study demonstrated that women suffering with postpartum psychosis do not show the normal increase in T cells following childbirth, but rather have an increase in monocytes (Bergink et al., 2013). Another study suggests widespread immune system changes, including replication of the findings demonstrating alterations in T cells and monocytes (Kumar et al., 2017). Additional evidence of monocyte involvement in postpartum psychosis comes from the evidence of altered microRNAs (miRNAs) in monocytes in patients with postpartum psychosis (Weigelt et al., 2013). MicroRNAs play a role in immune activation genes which have been shown to be altered in patients with postpartum psychosis (Bergink et al., 2013). Investigation into changes in miRNAs associated with postpartum psychosis identified downregulated miR-146a and miR-212 (Weigelt et al., 2013). Based on these provocative studies, it is tempting to speculate that miRNAs may be a useful biomarker for the risk of postpartum psychosis.

Interestingly, a two-hit model has been proposed in which inflammation or autoimmune conditions may increase vulnerability for psychotic disorders, and additional predisposing factors such as stress and endocrine changes could further alter immune system function and trigger psychosis (Bergink et al., 2014). In fact, the increase in monocytes in women with postpartum psychosis is associated with altered monocyte gene expression, including altered glucocorticoid gene expression (Bergink et al., 2013), suggesting a potential link between the hormonal and immunological factors contributing to postpartum psychosis. Given that postpartum period is already a vulnerable period for the manifestation of psychiatric disorders, it is feasible that the threshold to trigger psychosis is already lower in susceptible individuals.

Conclusions

This chapter summarizes the current state of knowledge regarding hormonal and inflammatory factors implicated in postpartum psychosis. The limited evidence presented also highlights the fact that this topic has not been extensively studied and the need for further investigation. Despite the limited number of studies, those groups that have explored this topic have elegantly demonstrated potential involvement of hormones, including estrogen, progesterone, oxytocin, prolactin, stress hormones (CRH, ACTH, and cortisol), and thyroid hormones, and immunological changes in postpartum psychosis. Much of the data on postpartum psychosis rely on correlational clinical studies, which are informative and provide the foundational basis for translationally relevant investigation; however, they are limited in their ability to provide additional mechanistic information regarding the underlying pathophysiology of disease. These studies are the building blocks, but we must build upon them in order to fully understand the neurobiological basis of postpartum psychosis.

References

Abel, K. M., Drake, R., & Goldstein, J. M. (2010). Sex differences in schizophrenia. *International Review of Psychiatry, 22*, 417–428.

Abou-Saleh, M. T., Ghubash, R., Karim, L., Krymski, M., & Bhai, I. (1998). Hormonal aspects of postpartum depression. *Psychoneuroendocrinology, 23*, 465–475.

Ahokas, A., & Aito, M. (1999). Role of estradiol in puerperal psychosis. *Psychopharmacology, 147*, 108–110.

Ahokas, A., Aito, M., & Rimon, R. (2000). Positive treatment effect of estradiol in postpartum psychosis: A pilot study. *The Journal of Clinical Psychiatry, 61*, 166–169.

Ahokas, A., Kaukoranta, J., Wahlbeck, K., & Aito, M. (2001). Estrogen deficiency in severe postpartum depression: Successful treatment with sublingual physiologic 17beta-estradiol: A preliminary study. *The Journal of Clinical Psychiatry, 62*, 332–336.

Akhondzadeh, S., Nejatisafa, A. A., Amini, H., Mohammadi, M. R., Larijani, B., Kashani, L., … Kamalipour, A. (2003). Adjunctive estrogen treatment in women with chronic schizophrenia: A double-blind, randomized, and placebo-controlled trial. *Progress in Neuro-Psychopharmacology and Biological Psychiatry, 27*, 1007–1012.

Bathla, M., Singh, M., & Relan, P. (2016). Prevalence of anxiety and depressive symptoms among patients with hypothyroidism. *Indian Journal of Endocrinology and Metabolism, 20*, 468–474.

Bell, A. F., Erickson, E. N., & Carter, C. S. (2014). Beyond labor: The role of natural and synthetic oxytocin in the transition to motherhood. *Journal of Midwifery and Women's Health, 59*, 35–42.

Bergemann, N., Mundt, C., Parzer, P., Jannakos, I., Nagl, I., Salbach, B., … Resch, F. (2005). Plasma concentrations of estradiol in women suffering from schizophrenia treated with conventional versus atypical antipsychotics. *Schizophrenia Research, 73*, 357–366.

Bergink, V., Burgerhout, K. M., Weigelt, K., Pop, V. J., de Wit, H., Drexhage, R. C., … Drexhage, H. A. (2013). Immune system dysregulation in first-onset postpartum psychosis. *Biological Psychiatry, 73*, 1000–1007.

Bergink, V., Gibney, S. M., & Drexhage, H. A. (2014). Autoimmunity, inflammation, and psychosis: A search for peripheral markers. *Biological Psychiatry, 75*, 324–331.

Bergink, V., Kushner, S. A., Pop, V., Kuijpens, H., Lambregtse-van den Berg, M. P., Drexhage, R. C., … Drexhage, H. A. (2011). Prevalence of autoimmune thyroid dysfunction in postpartum psychosis. *The British Journal of Psychiatry: The Journal of Mental Science, 198*, 264–268.

Bergink, V., Pop, V. J. M., Nielsen, P. R., Agerbo, E., Munk-Olsen, T., & Liu, X. (2018). Comorbidity of autoimmune thyroid disorders and psychiatric disorders during the postpartum period: A Danish nationwide register-based cohort study. *Psychological Medicine, 48*, 1291–1298.

Bergink, V., Rasgon, N., & Wisner, K. L. (2016). Postpartum psychosis: Madness, mania, and melancholia in motherhood. *American Journal of Psychiatry, 173*, 1179–1188.

Blackmore, E. R., Moynihan, J. A., Rubinow, D. R., Pressman, E. K., Gilchrist, M., & O'Connor, T. G. (2011). Psychiatric symptoms and proinflammatory cytokines in pregnancy. *Psychosomatic Medicine, 73*, 656–663.

Bloch, M., Daly, R. C., & Rubinow, D. R. (2003). Endocrine factors in the etiology of postpartum depression. *Comprehensive Psychiatry, 44*, 234–246.

Bloch, M., Rubinow, D. R., Schmidt, P. J., Lotsikas, A., Chrousos, G. P., & Cizza, G. (2005). Cortisol response to ovine corticotropin-releasing hormone in a model of pregnancy and parturition in Euthymic women with and without a history of postpartum depression. *The Journal of Clinical Endocrinology and Metabolism, 90*, 695–699.

Bloch, M., Schmidt, P. J., Danaceau, M., Murphy, J., Nieman, L., & Rubinow, D. R. (2000). Effects of gonadal steroids in women with a history of postpartum depression. *American Journal of Psychiatry, 157*, 924–930.

Bokhari, R., Bhatara, V. S., Bandettini, F., & Michael McMillin, J. (1998). Postpartum psychosis and postpartum Thyroiditis1 presented at the 27th congress of International Society of Psychoneuroendocrinology (August 1996), Cascais, Portugal.1. *Psychoneuroendocrinology, 23*, 643–650.

Boufidou, F., Lambrinoudaki, I., Argeitis, J., Zervas, I. M., Pliatsika, P., Leonardou, A. A., ... Nikolaou, C. (2009). CSF and plasma cytokines at delivery and postpartum mood disturbances. *Journal of Affective Disorders, 115*, 287–292.

Brummelte, S., & Galea, L. A. M. (2010). Depression during pregnancy and postpartum: Contribution of stress and ovarian hormones. *Progress in Neuro-Psychopharmacology and Biological Psychiatry, 34*, 766–776.

Brunton, P. J., & Russell, J. A. (2008). Attenuated hypothalamo-pituitary-adrenal axis responses to immune challenge during pregnancy: The neurosteroid–opioid connection. *The Journal of Physiology, 586*, 369–375.

Brunton, P. J., & Russell, J. A. (2011). Allopregnanolone and suppressed hypothalamo–pituitary–adrenal axis stress responses in late pregnancy in the rat. *Stress: The International Journal on the Biology of Stress, 14*, 6–12.

Brunton, P. J., Russell, J. A., & Douglas, A. J. (2008). Adaptive responses of the maternal hypothalamic-pituitary-adrenal axis during pregnancy and lactation. *Journal of Neuroendocrinology, 20*, 764–776.

Capuron, L., & Miller, A. H. (2004). Cytokines and psychopathology: Lessons from interferon-α. *Biological Psychiatry, 56*, 819–824.

Cassidy-Bushrow, A. E., Peters, R. M., Johnson, D. A., & Templin, T. N. (2012). Association of depressive symptoms with inflammatory biomarkers among pregnant African-American women. *Journal of Reproductive Immunology, 94*, 202–209.

Chang, S. S., & Renshaw, D. C. (1986). Psychosis and pregnancy. *Comprehensive Therapy, 12*, 36–41.

Chatterjee, O., Patil, K., Sahu, A., Gopalakrishnan, L., Mol, P., Advani, J., ... Prasad, T. S. K. (2016). An overview of the oxytocin-oxytocin receptor signaling network. *Journal of Cell Communication and Signaling, 10*, 355–360.

Corcoran, C., Walker, E., Huot, R., Mittal, V., Tessner, K., Kestler, L., & Malaspina, D. (2003). The stress cascade and schizophrenia: Etiology and onset. *Schizophrenia Bulletin, 29*, 671–692.

Corwin, E. J., & Pajer, K. (2008). The psychoneuroimmunology of postpartum depression. *Journal of Women's Health (2002), 17*, 1529–1534.

Dalton, K. (1959). Menstruation and acute psychiatric illnesses. *British Medical Journal, 1*, 148–149.

Dantzer, R., O'Connor, J. C., Freund, G. G., Johnson, R. W., & Kelley, K. W. (2008). From inflammation to sickness and depression: When the immune system subjugates the brain. *Nature Reviews. Neuroscience, 9*, 46–56.

De Weerth, C., & Buitelaar, J. K. (2005). Cortisol awakening response in pregnant women. *Psychoneuroendocrinology, 30*, 902–907.

Dowlati, Y., Herrmann, N., Swardfager, W., Liu, H., Sham, L., Reim, E. K., & Lanctôt, K. L. (2010). A meta-analysis of cytokines in major depression. *Biological Psychiatry, 67*, 446–457.

Dubovsky, A. N., Arvikar, S., Stern, T. A., & Axelrod, L. (2012). The neuropsychiatric complications of glucocorticoid use: Steroid psychosis revisited. *Psychosomatics, 53*, 103–115.

Felger, J. C., & Lotrich, F. E. (2013). Inflammatory cytokines in depression: Neurobiological mechanisms and therapeutic implications. *Neuroscience, 246*, 199–229.

Fink, G., Sumner, B. E., Rosie, R., Grace, O., & Quinn, J. P. (1996). Estrogen control of central neurotransmission: Effect on mood, mental state, and memory. *Cellular and Molecular Neurobiology, 16*, 325–344.

Garg, D., Ng, S. S. M., Baig, K. M., Driggers, P., & Segars, J. (2017). Progesterone-mediated non-classical signaling. *Trends in Endocrinology and Metabolism, 28*, 656–668.

Garner, B., Pariante, C. M., Wood, S. J., Velakoulis, D., Phillips, L., Soulsby, B., … Pantelis, C. (2005). Pituitary volume predicts future transition to psychosis in individuals at ultra-high risk of developing psychosis. *Biological Psychiatry, 58*, 417–423.

Gimeno, D., Kivimäki, M., Brunner, E. J., Elovainio, M., De Vogli, R., Steptoe, A., … Ferrie, J. E. (2009). Associations of C-reactive protein and interleukin-6 with cognitive symptoms of depression: 12-year follow-up of the Whitehall II study. *Psychological Medicine, 39*, 413–423.

Girshkin, L., Matheson, S. L., Shepherd, A. M., & Green, M. J. (2014). Morning cortisol levels in schizophrenia and bipolar disorder: A meta-analysis. *Psychoneuroendocrinology, 49*, 187–206.

Gleicher, N. (2007). Postpartum depression, an autoimmune disease? *Autoimmunity Reviews, 6*, 572–576.

Glynn, L. M., Davis, E. P., & Sandman, C. A. (2013). New insights into the role of perinatal HPA-axis dysregulation in postpartum depression. *Neuropeptides, 47*, 363–370.

Gogos, A., Sbisa, A. M., Sun, J., Gibbons, A., Udawela, M., & Dean, B. (2015). A role for estrogen in schizophrenia: Clinical and preclinical findings. *International Journal of Endocrinology, 2015*, 615356.

Gregoire, A. J. P., Kumar, R., Everitt, B., & Studd, J. W. W. (1996). Transdermal oestrogen for treatment of severe postnatal depression. *The Lancet, 347*, 930–933.

Groer, M. W., & Morgan, K. (2007). Immune, health and endocrine characteristics of depressed postpartum mothers. *Psychoneuroendocrinology, 32*, 133–139.

Grosse, L., Hoogenboezem, T., Ambrée, O., Bellingrath, S., Jörgens, S., de Wit, H. J., … Drexhage, H. A. (2016). Deficiencies of the T and natural killer cell system in major depressive disorder: T regulatory cell defects are associated with inflammatory monocyte activation. *Brain, Behavior, and Immunity, 54*, 38–44.

Guintivano, J., Sullivan, P. F., Stuebe, A. M., Penders, T., Thorp, J., Rubinow, D. R., & Meltzer-Brody, S. (2018). Adverse life events, psychiatric history, and biological predictors of postpartum depression in an ethnically diverse sample of postpartum women. *Psychological Medicine, 48*, 1190–1200.

Harris, B., Johns, S., Fung, H., Thomas, R., Walker, R., Read, G., & Riad-Fahmy, D. (1989). The hormonal environment of post-natal depression. *The British Journal of Psychiatry: The Journal of Mental Science, 154*, 660–667.

Harvey, C. B., & Williams, G. R. (2002). Mechanism of thyroid hormone action. *Thyroid, 12*, 441–446.

Howren, M. B., Lamkin, D. M., & Suls, J. (2009). Associations of depression with C-reactive protein, IL-1, and IL-6: A meta-analysis. *Psychosomatic Medicine, 71*, 171–186.

Huber, T. J., Rollnik, J., Wilhelms, J., von zur Mühlen, A., Emrich, H. M., & Schneider, U. (2001). Estradiol levels in psychotic disorders. *Psychoneuroendocrinology, 26*, 27–35.

Hucklebridge, F. H., Smith, M. D., Clow, A., Evans, P., Glover, V., Taylor, A., … Lydyard, P. M. (1994). Dysphoria and immune status in postpartum women. *Biological Psychology, 37*, 199–206.

Kanes, S., Colquhoun, H., Gunduz-Bruce, H., Raines, S., Arnold, R., Schacterle, A., … Meltzer-Brody, S. (2017). Brexanolone (SAGE-547 injection) in post-partum depression: A randomised controlled trial. *The Lancet, 390*, 480–489.

Kanes, S. J., Colquhoun, H., Doherty, J., Raines, S., Hoffmann, E., Rubinow, D. R., & Meltzer-Brody, S. (2017). Open-label, proof-of-concept study of brexanolone in the treatment of severe postpartum depression. *Human Psychopharmacology, 32*, e2576.

Kendell, R. E., Chalmers, J. C., & Platz, C. (1987). Epidemiology of puerperal psychoses. *British Journal of Psychiatry, 150*, 662–673.

Kroll-Desrosiers, A. R., Nephew, B. C., Babb, J. A., Guilarte-Walker, Y., Moore Simas, T. A., & Deligiannidis, K. M. (2017). Association of peripartum synthetic oxytocin administration and depressive and anxiety disorders within the first postpartum year. *Depression and Anxiety, 34*, 137–146.

Kulkarni, J., Gavrilidis, E., Wang, W., Worsley, R., Fitzgerald, P. B., Gurvich, C., … Burger, H. (2014). Estradiol for treatment-resistant schizophrenia: A large-scale randomized-controlled trial in women of child-bearing age. *Molecular Psychiatry, 20*, 695.

Kulkarni, J., Riedel, A., de Castella, A. R., Fitzgerald, P. B., Rolfe, T. J., Taffe, J., & Burger, H. (2001). Estrogen — a potential treatment for schizophrenia. *Schizophrenia Research, 48*, 137–144.

Kumar, M. M., Venkataswamy, M. M., Sathyanarayanan, G., Thippeswamy, H., Chandra, P. S., & Mani, R. S. (2017). Immune system aberrations in postpartum psychosis: An immunophenotyping study from a tertiary care neuropsychiatric hospital in India. *Journal of Neuroimmunology, 310*, 8–13.

Lawrie, T. A., Herxheimer, A., & Dalton, K. (2000). Oestrogens and progestogens for preventing and treating postnatal depression. *Cochrane Database of Systematic Reviews*, Cd001690.

Levine, J., Barak, Y., Chengappa, K. N. R., Rapoport, A., Rebey, M., & Barak, V. (1999). Cerebrospinal cytokine levels in patients with acute depression. *Neuropsychobiology, 40*, 171–176.

Licinio, J. (2009). Potential diagnostic markers for postpartum depression point out to altered immune signaling. *Molecular Psychiatry, 15*, 1.

Luppi, P. (2003). How immune mechanisms are affected by pregnancy. *Vaccine, 21*, 3352–3357.

MacKenzie, G., & Maguire, J. (2013). Neurosteroids and GABAergic signaling in health and disease. *Biomolecular Concepts, 4*, 29–42.

Magiakou, M. A., Mastorakos, G., Rabin, D., Dubbert, B., Gold, P. W., & Chrousos, G. P. (1996). Hypothalamic corticotropin-releasing hormone suppression during the postpartum period: Implications for the increase in psychiatric manifestations at this time. *The Journal of Clinical Endocrinology and Metabolism, 81*, 1912–1917.

Mah, B. L., Van Ijzendoorn, M. H., Out, D., Smith, R., & Bakermans-Kranenburg, M. J. (2017). The effects of intranasal oxytocin administration on sensitive caregiving in mothers with postnatal depression. *Child Psychiatry and Human Development, 48*, 308–315.

Marx, C. E., Bradford, D. W., Hamer, R. M., Naylor, J. C., Allen, T. B., Lieberman, J. A., … Kilts, J. D. (2011). Pregnenolone as a novel therapeutic candidate in schizophrenia: Emerging preclinical and clinical evidence. *Neuroscience, 191*, 78–90.

Marx, C. E., Stevens, R. D., Shampine, L. J., Uzunova, V., Trost, W. T., Butterfield, M. I., … Lieberman, J. A. (2005). Neuroactive steroids are altered in schizophrenia and bipolar disorder: Relevance to pathophysiology and therapeutics. *Neuropsychopharmacology, 31*, 1249.

McNeil, T. F. (1987). A prospective study of postpartum psychoses in a high-risk group. *Acta Psychiatrica Scandinavica, 75*, 35–43.

Meltzer-Brody, S., Kanes, S., Riesenberg, R., Rubinow, D., Maximos, B., & Colquhoun, H. (2018). Phase 3 study evaluating brexanolone, a GABAA receptor modulator, in severe postpartum depression [29B]. *Obstetrics and Gynecology, 131*, 27S.

Miller, A. H., Haroon, E., Raison, C. L., & Felger, J. C. (2013). Cytokine targets in the brain: Impact on neurotransmitters and neurocircuits. *Depression and Anxiety, 30*, 297–306.

Mittal, V. A., Dhruv, S., Tessner, K. D., Walder, D. J., & Walker, E. F. (2007). The relations among putative biorisk markers in schizotypal adolescents: Minor physical anomalies, movement abnormalities, and salivary cortisol. *Biological Psychiatry, 61*, 1179–1186.

Mor, G., & Cardenas, I. (2010). The immune system in pregnancy: A unique complexity. *American Journal of Reproductive Immunology (New York, N.Y.: 1989), 63*, 425–433.

Mor, G., Cardenas, I., Abrahams, V., & Guller, S. (2011). Inflammation and pregnancy: The role of the immune system at the implantation site. *Annals of the New York Academy of Sciences, 1221*, 80–87.

Moses-Kolko, E. L., Berga, S. L., Kalro, B., Sit, D. K. Y., & Wisner, K. L. (2009). Transdermal estradiol for postpartum depression: A promising treatment option. *Clinical Obstetrics and Gynecology, 52*, 516–529.

Müller, N., Weidinger, E., Leitner, B., & Schwarz, M. J. (2015). The role of inflammation in schizophrenia. *Frontiers in Neuroscience, 9*, 372.

Nonacs, R., & Cohen, L. S. (1998). Postpartum mood disorders: Diagnosis and treatment guidelines. *The Journal of clinical psychiatry, 59*(Suppl. 2), 34–40.

Oakley, R. H., & Cidlowski, J. A. (2013). The biology of the glucocorticoid receptor: New signaling mechanisms in health and disease. *The Journal of Allergy and Clinical Immunology, 132*, 1033–1044.

Osborne, L. M., Gispen, F., Sanyal, A., Yenokyan, G., Meilman, S., & Payne, J. L. (2017). Lower allopregnanolone during pregnancy predicts postpartum depression: An exploratory study. *Psychoneuroendocrinology, 79*, 116–121.

Ostiguy, C. S., Ellenbogen, M. A., Walker, C. D., Walker, E. F., & Hodgins, S. (2011). Sensitivity to stress among the offspring of parents with bipolar disorder: A study of daytime cortisol levels. *Psychological Medicine, 41*, 2447–2457.

Pariante, C. M. (2008). Pituitary volume in psychosis: The first review of the evidence. *Journal of Psychopharmacology, 22*, 76–81.

Pariante, C. M. (2009). Risk factors for development of depression and psychosis: Glucocorticoid receptors and pituitary implications for treatment with antidepressant and glucocorticoids. *Annals of the New York Academy of Sciences, 1179*, 144–152.

Pariante, C. M., & Lightman, S. L. (2008). The HPA axis in major depression: Classical theories and new developments. *Trends in Neurosciences, 31*, 464–468.

Perlman, W. R., Webster, M. J., Kleinman, J. E., & Weickert, C. S. (2004). Reduced glucocorticoid and estrogen receptor alpha messenger ribonucleic acid levels in the amygdala of patients with major mental illness. *Biological Psychiatry, 56*, 844–852.

Pollak, Y., & Yirmiya, R. (2002). Cytokine-induced changes in mood and behaviour: Implications for 'depression due to a general medical condition', immunotherapy and antidepressive treatment. *International Journal of Neuropsychopharmacology, 5*, 389–399.

Radhakrishnan, A., Raju, R., Tuladhar, N., Subbannayya, T., Thomas, J. K., Goel, R., … Chatterjee, A. (2012). A pathway map of prolactin signaling. *Journal of Cell Communication and Signaling, 6*, 169–173.

Raison, C. L., Capuron, L., & Miller, A. H. (2006). Cytokines sing the blues: Inflammation and the pathogenesis of depression. *Trends in Immunology, 27*, 24–31.

Rajkumar, R. P. (2014). Prolactin and psychopathology in schizophrenia: A literature review and reappraisal. *Schizophrenia Research and Treatment, 2014*, 12.

Rich, M. E., & Caldwell, H. K. (2015). A role for oxytocin in the etiology and treatment of schizophrenia. *Frontiers in Endocrinology, 6*, 90.

Santos, N. C., Costa, P., Ruano, D., Macedo, A., Soares, M. J., Valente, J., … Palha, J. A. (2012). Revisiting thyroid hormones in schizophrenia. *Journal of Thyroid Research, 2012*, 569147.

Schmidt, H. J. (1943). The use of progesterone in the treatment of postpartum psychosis. *Journal of the American Medical Association, 121*, 190–192.
Schüle, C., Nothdurfter, C., & Rupprecht, R. (2014). The role of allopregnanolone in depression and anxiety. *Progress in Neurobiology, 113*, 79–87.
Schulte, H. M., Weisner, D., & Allolio, B. (1990). The corticotrophin releasing hormone test in late pregnancy: Lack of adrenocorticotrophin and cortisol response. *Clinical Endocrinology, 33*, 99–106.
Shilling, P. D., & Feifel, D. (2016). Potential of oxytocin in the treatment of schizophrenia. *CNS Drugs, 30*, 193–208.
Sichel, D. A., Cohen, L. S., Robertson, L. M., Ruttenberg, A., & Rosenbaum, J. F. (1995). Prophylactic estrogen in recurrent postpartum affective disorder. *Biological Psychiatry, 38*, 814–818.
Sit, D., Rothschild, A. J., & Wisner, K. L. (2006). A review of postpartum psychosis. *Journal of Women's Health, 15*, 352–368.
Skrundz, M., Bolten, M., Nast, I., Hellhammer, D. H., & Meinlschmidt, G. (2011). Plasma oxytocin concentration during pregnancy is associated with development of postpartum depression. *Neuropsychopharmacology, 36*, 1886–1893.
Soldin, O. P., Guo, T., Weiderpass, E., Tractenberg, R. E., Hilakivi-Clarke, L., & Soldin, S. J. (2005). Steroid hormone levels in pregnancy and 1 year postpartum using isotope dilution tandem mass spectrometry. *Fertility and Sterility, 84*, 701–710.
Spinelli, M. G. (2009). Postpartum psychosis: Detection of risk and management. *American Journal of Psychiatry, 166*, 405–408.
Stuebe, A. M., Grewen, K., & Meltzer-Brody, S. (2013). Association between maternal mood and oxytocin response to breastfeeding. *Journal of Women's Health, 22*, 352–361.
Stuebe, A. M., Grewen, K., Pedersen, C. A., Propper, C., & Meltzer-Brody, S. (2012). Failed lactation and perinatal depression: Common problems with shared neuroendocrine mechanisms? *Journal of Women's Health, 21*, 264–272.
Sun, J., Walker, A. J., Dean, B., van den Buuse, M., & Gogos, A. (2016). Progesterone: The neglected hormone in schizophrenia? A focus on progesterone-dopamine interactions. *Psychoneuroendocrinology, 74*, 126–140.
Tal, R., Taylor, H. S., Burney, R. O., Mooney, S. B., & Giudice, L. C. (2000). Endocrinology of pregnancy. In L. J. De Groot, G. Chrousos, K. Dungan, K. R. Feingold, A. Grossman, J. M. Hershman, et al.A. Vinik (Eds.), *Endotext*. South Dartmouth (MA): MDText.com, Inc.
Tay, T. L., Béchade, C., D'Andrea, I., St-Pierre, M.-K., Henry, M. S., Roumier, A., & Tremblay, M.-E. (2017). Microglia gone rogue: Impacts on psychiatric disorders across the lifespan. *Frontiers in Molecular Neuroscience, 10*, 421.
VanderKruik, R., Barreix, M., Chou, D., Allen, T., Say, L., Cohen, L. S., ... Maternal Morbidity Working Group. (2017). The global prevalence of postpartum psychosis: A systematic review. *BMC Psychiatry, 17*, 272.
Walker, E. F., Trotman, H., Pearce, B. D., Addington, J., Cadenhead, K. S., Cornblatt, B. A., ... Woods, S. W. (2013). Cortisol levels and risk for psychosis: Initial findings from the North American Prodrome Longitudinal study. *Biological Psychiatry, 74*, 410–417.
Walsh, B., Seidman, S. N., Sysko, R., & Gould, M. (2002). Placebo response in studies of major depression: Variable, substantial, and growing. *Journal of the American Medical Association, 287*, 1840–1847.
Weigelt, K., Bergink, V., Burgerhout, K. M., Pescatori, M., Wijkhuijs, A., & Drexhage, H. A. (2013). Down-regulation of inflammation-protective microRNAs 146a and 212 in monocytes of patients with postpartum psychosis. *Brain, Behavior, and Immunity, 29*, 147–155.

Yehuda, R., Boisoneau, D., Mason, J. W., & Giller, E. L. (1993). Glucocorticoid receptor number and cortisol excretion in mood, anxiety, and psychotic disorders. *Biological Psychiatry, 34*, 18–25.

Zhang, X. Y., Zhou, D. F., Cao, L. Y., Wu, G. Y., & Shen, Y. C. (2005). Cortisol and cytokines in chronic and treatment-resistant patients with schizophrenia: Association with psychopathology and response to antipsychotics. *Neuropsychopharmacology, 30*, 1532.

Zunszain, P. A., Hepgul, N., & Pariante, C. M. (2013). Inflammation and depression. *Cuurent Topics in Behavioral Neurosciences, 14*, 135–151.

CHAPTER 13

Functional MRI biomarkers of peripartum psychiatric disorders

Megan Hare[1], Christy Duan[2], Kristina M. Deligiannidis[3,4,5]
[1]Department of Psychology, Florida International University, Miami, FL, United States of America; [2]Department of Psychiatry, Zucker Hillside Hospital, Northwell Health, Glen Oaks, NY, United States of America; [3]Women's Behavioral Health, Department of Psychiatry, Zucker Hillside Hospital, Northwell Health, Glen Oaks, NY, United States of America; [4]Departments of Psychiatry and Obstetrics & Gynecology, Zucker School of Medicine at Hofstra/Northwell, Hempstead, NY, United States of America; [5]Feinstein Institute for Medical Research, Manhasset, NY, United States of America

Introduction

Peripartum psychiatric disorders affect 10%–20% of women (Fisher et al., 2012), representing a significant public health problem (Bauer, Knapp, & Parsonage, 2016). These disorders, including depression, psychosis, and anxiety, can have substantial adverse effects on the well-being of the mother and child with lasting consequences. Psychiatric symptoms during pregnancy are associated with an increased risk for maternal substance abuse (Judd et al., 2014), preeclampsia (Judd et al., 2014), preterm delivery, and low infant birth weight (Grote et al., 2010). Postpartum psychiatric illness can impair a woman's ability to care for herself and infant, negatively affecting a child's cognitive, behavioral, and emotional development (Goodman et al., 2011, O'Connor, Heron, Golding, Beveridge, & Glover, 2002). In severe circumstances, child abuse (Plant, Pariante, Sharp, & Pawlby, 2015), infanticide (Spinelli, 2004), and maternal suicide (Draper et al., 2018; Lindahl, Pearson, & Colpe, 2005) can occur.

While the *Diagnostic and Statistical Manual of Mental Disorders* (DSM-5) has specifiers for some disorders occurring in the peripartum period (i.e., unipolar depression, bipolar disorder, psychotic disorder), others including anxiety disorders and obsessive compulsive disorder (OCD) are not acknowledged (APA, 2013). The field is beginning to move in the right direction, as recent guidelines by the American Psychiatric Association (APA) recommend screening all perinatal women for depression, anxiety, and psychotic disorders (Byatt et al., 2018). This is vital as comorbid psychiatric disorders are common (Dindo, Elmore, O'Hara, & Stuart, 2017) and associated with higher symptom severity (Hirschfeld, 2001), increased treatment resistance (Sharma, Khan, Corpse, & Sharma, 2008), and

Biomarkers of Postpartum Psychiatric Disorders
ISBN 978-0-12-815508-0
https://doi.org/10.1016/B978-0-12-815508-0.00013-8

suicidality (Newport, Levey, Pennell, Ragan, & Stowe, 2007). Accurately detecting and diagnosing comorbidity is essential, as it may influence treatment planning. Yet disorders in the peripartum are frequently underdetected (Sharma et al., 2008) and undertreated (Vesga-Lopez et al., 2008), leaving women with prolonged symptoms and significant impairment (Hung, Liu, & Yang, 2017). Earlier and more accurate diagnosis could be achieved with increased understanding of the pathophysiology of these disorders.

Research into the underpinnings of peripartum psychiatric illness has increased (Jones et al., 2007; Kroll-Desrosiers et al., 2017; Magiakou et al., 1996; Parry et al., 2008), yet the underlying neurocircuitry is still poorly understood. With consistent evidence that psychiatric illnesses are brain disorders associated with complex, multimodal patterns of brain abnormalities, the need for future research into the neural underpinning of these disorders is paramount. Increasing our understanding of peripartum psychiatric illness at the neurobiological level, with the objective of establishing functional biomarkers, will aid us in effectively detecting, diagnosing, and treating these disorders in the antepartum and postpartum periods. This chapter will review the current functional neuroimaging literature on peripartum depression (unipolar and bipolar), psychosis, anxiety (generalized anxiety disorder and posttraumatic stress disorder), and OCD and identify next steps in advancing this nascent literature. Table 13.1 indicates abbreviations used in this chapter.

Methods

Papers were searched on MEDLINE, PsychINFO, Web of Science, Scopus, Embase, and PubMed with the following key words: ("pregnancy" or "antepartum" or "perinatal" or "motherhood" or "postpartum" or "maternal" or "antenatal" or "postnatal" or "prepartum" or "peripartum" or "mothers") AND ("fMRI" or "functional neuroimaging" or "magnetic resonance imaging" or "MRI") AND ("depression" or "depressive" or "psychosis" or "anxiety" or "GAD" or "generalized anxiety disorder" or "OCD" or "obsessive compulsive disorder" or "stress" or "PTSD" or "post-traumatic stress disorder" or "bipolar" or "bipolar disorder"). This review is limited to papers published in the last 10 years, in English, and on subjects up to 6 months postpartum. Additional articles were identified by reviewing bibliographies of review articles identified within the literature search. Case reports were excluded.

Table 13.1 Abbreviations used in this chapter.

DSM-5	Diagnostic and statistical manual of mental disorders
fMRI	Functional magnetic resonance imaging
BOLD	Blood-oxygenation level dependent
RSFC	Resting-state functional connectivity
DMN	Default mode network
VMPFC	Ventromedial prefrontal cortex
PCC	Posterior cingulate cortex
IPL	Inferior parietal lobule
CEN	Central executive network
DLPFC	Dorsolateral prefrontal cortex
DMPFC	Dorsomedial prefrontal cortex
SN	Salience network
IC	Insula cortex
ACC	Anterior cingulate cortex
MFG	Middle frontal gyrus
PPD	Postpartum depression
MDD	Major depressive disorder
DAN	Dorsal attention network
FN	Frontoparietal network
STG	Superior temporal gyrus
CHR	Clinical high risk
OCD	Obsessive–compulsive disorder
GAD	Generalized anxiety disorder
PTSD	Posttraumatic stress disorder
EPDS	Edinburgh postnatal depression scale

Results

An overview of functional magnetic resonance imaging

Functional magnetic resonance imaging (fMRI) estimates brain activity through the measurement of cerebral blood-oxygenation-level dependent (BOLD) changes (Logothetis, Pauls, Augath, Trinath, & Oeltermann, 2001). Brain activity patterns can be measured at rest or while completing a task during the scan. The pattern of activity implies "connectivity," the concept that separate brain regions communicate with each other in specific networks (van den Heuvel & Hulshoff Pol, 2010). A network is a consistent group of spatially distinct brain regions that activate over time to accomplish a common function. There is no consensus about how many networks there are, or which specific brain regions make up a network, but several networks have been implicated in psychiatric disorders. The default mode network (DMN), a diffuse network of brain regions most active at rest, is involved in internal mentation and external environmental monitoring. Regions frequently implicated include the ventromedial prefrontal cortex (VMPFC) and posterior cingulate cortex (PCC) (Seeley et al., 2007). Other studies suggest that the precuneus and inferior parietal lobule (IPL) are also involved (Buckner, Andrews-Hanna, & Schacter, 2008).The central executive network (CEN) regulates attention and higher-order cognitive functions (Seeley et al., 2007). It includes the dorsolateral prefrontal cortex (DLPFC), dorsomedial prefrontal cortex (DMPFC), and PCC (Bressler & Menon, 2010; Sridharan, Levitin, & Menon, 2008). The salience network (SN) assesses the salience of internal and external stimuli (Newport et al., 2007; Seeley et al., 2007). The literature suggests that the SN modulates DMN activation for internal cognitive processes and CEN activation for external cognitive processes (Borst & Anderson, 2013; Sridharan et al., 2008). This network begins in the anterior insula cortex (IC) and dorsal anterior cingulate cortex (ACC) before connecting with subcortical and limbic brain regions involved in reward and motivation (Sridharan et al., 2008). Resting-state functional connectivity (RSFC) studies examine how brain networks are functionally connected in conditions of health or disease while subjects are at rest (van den Heuvel & Hulshoff Pol, 2010). In seed-based RSFC studies, signal from a specific voxel or cluster of voxels (i.e., regions of interest or seeds) is used to calculate correlations with other voxels of the brain, allowing for a more accurate picture of specific connectivity in the brain areas of interest (Biswal, Yetkin, Haughton, & Hyde, 1995; Smitha, Arun, Rajesh, Thomas, & Kesavadas, 2017). While RSFC studies allow us to investigate the overall functional organization of brain networks, task-based fMRI studies allow us to identify networks that are involved while a subject is involved in an active task.

fMRI studies in healthy postpartum and nonpostpartum women

The peripartum period is marked by complex physiological changes affecting mood and cognitive functioning (Buckwalter et al., 1999; Henry & Sherwin, 2012). Understanding brain function during pregnancy requires comparison with nonpregnant and postpartum states. There is one RSFC and two task-based studies that compared healthy nonpostpartum and postpartum women. The RSFC study examined neural activity related to cognitive impairment in 23 healthy postpartum women and 23 healthy nonpostpartum women (Zheng et al., 2018). Postpartum women showed significantly decreased spontaneous neural activity in the left PCC, right superior frontal gyrus (SFG), and right middle frontal gyrus (MFG) compared to the healthy nonpostpartum women.

A task-based fMRI study examined executive function in 13 healthy postpartum women and 13 healthy nonpostpartum women using the response inhibition Go/No-Go task (Bannbers et al., 2013). Healthy postpartum women performed the fMRI task at 48 h and 4–7 weeks postpartum, while healthy nonpostpartum women performed the same task in late luteal phase, a high progesterone period, and mid-follicular phase, a low progesterone period. Postpartum women had decreased prefrontal activity overall compared to nonpostpartum women when engaging in the same task. Furthermore, late postpartum women (4–7 weeks) had even greater decreases in prefrontal activity than early postpartum women (48 h). No regions of interest, including the inferior frontal gyri, anterior cingulate, precentral gyri, were correlated with serum concentrations of ovarian steroids or neurosteroids (Bannbers et al., 2013).

Using the same sample of women as the study above, a second task-based fMRI study was conducted examining longitudinal changes in emotional processing by comparing 13 healthy postpartum women within 48 h and 4–6 weeks of delivery to 15 healthy nonpostpartum women at the same time points in the menstrual cycle (Gingnell et al., 2015). Women completed an emotion processing task where they matched angry and fearful facial expressions, as well as a simple sensorimotor control task where they matched shapes (Hariri et al., 2002). Brain reactivity was not correlated with serum estradiol or progesterone levels. When compared to nonpregnant women, postpartum women had greater reactivity in insula and inferior frontal gyrus. When compared to early postpartum women, late postpartum women had higher reactivity in the right insula, left MFG, and bilateral inferior frontal gyrus (Gingnell et al., 2015).

Functional connectivity literature comparing healthy nonpostpartum women to healthy postpartum women is limited. There are no studies comparing prepregnancy to pregnancy to postpartum brain changes. Hence, there is very little on the physiologic changes the brain goes through during this time of heightened risk for development of psychiatric illness. No concrete conclusions can be drawn about what connectivity changes, if any, occur across the peripartum period.

Unipolar peripartum depression

The most studied peripartum psychiatric disorder is unipolar depression, experienced by 16% of women during pregnancy and 20% postpartum (Woody, Ferrari, Siskind, Whiteford, & Harris, 2017). Consistently, research shows that antepartum and postpartum depression negatively and differentially affect child development (Goodman et al., 2011), with exposure to maternal depressive symptoms as one of the most well-established risk factors for the development of child psychopathology (Kingston et al., 2018). While symptoms can be similar to major depressive disorder (MDD) outside the peripartum period, peripartum depression may have a different clinical presentation (Hendrick, Altshuler, Strouse, & Grosser, 2000; Putnam et al., 2017) and rate of comorbid psychiatric disorders (Dindo et al., 2017). Additionally, 40% of women will have their first episode of depression during the peripartum period, indicating that peripartum depression is not always just a continuation of MDD (Wisner et al., 2013).

Peripartum depression is a heterogeneous disorder with distinct phenotypes differing in symptom severity, timing of symptom onset, history of previous psychiatric disorder, pregnancy complications, and presence of suicidal ideation (Kingston et al., 2018; Putnam et al., 2015; Putnam et al., 2017). The DSM-5 (APA, 2013) uses a specifier of "peripartum onset" with a diagnosis of MDD but does not distinguish between antenatal and postpartum onset. Although fMRI research in PPD is still in its infancy, researchers have begun to examine differences in neurocircuitry in comparison to healthy postpartum women and women with MDD. In the future, functional neuroimaging may give researchers insight into the biological signature of each subtype, facilitating more accurate diagnosis and treatment.

fMRI studies in unipolar peripartum depression

Networks of increasing interest in nonperipartum depression include the DMN, CEN, the SN, dorsal attention network (DAN), and frontoparietal network (FN) (Kaiser, Andrews-Hanna, Wager, & Pizzagalli, 2015). PPD research has begun to focus on these networks; however, there are still critical research gaps. We identified nine functional neuroimaging studies in the PPD literature: three RSFC and six task-based.

One brain region of significant interest in both nonperipartum and peripartum depression is the amygdala, located in the SN. This key brain structure plays a critical role in the processing of memory, decision-making, emotional processing, and maternal behavior (Barrett & Fleming, 2011). Research has found a significant relationship between depressive symptoms and amygdala functioning in women with and without PPD. In a seed-based RSFC study within 9 weeks postpartum, women with PPD had reduced functional connectivity between bilateral amygdalae, ACC, and bilateral DLPFC and between the left DLPFC and right amygdala. Furthermore, women with PPD had weaker connectivity between the right hippocampus (located in the DMN network) and right DLPFC, when compared to healthy postpartum women (Deligiannidis et al., 2013). This is the first study reporting decreased connectivity among three separate brain networks, the DMN, SN, and CEN, in PPD. These results are consistent with previous RSFC studies in nonperipartum depression showing decreased functional connectivity between cortical and limbic structures (Anand et al., 2005). A larger follow-up RSFC study by the same research group in women diagnosed with PPD and healthy postpartum women used an Independent Component Analysis approach to identify network connectivity differences between groups (Deligiannidis et al., 2019). Within the DMN, an area of the DMPFC had greater connectivity with the rest of the DMN and was correlated to depression scores and plasma allopregnanolone concentrations. A subsequent seed-based analysis of the DMPFC region showed reduced RSFC between the DMPFC and the precuneus, PDD and supramarginal gyrus/angular gyrus in women diagnosed with PPD (Deligiannidis et al., 2019). The precuneus is structurally connected to the PCC, the ACC, thalamus, and striatum (Cavanna & Trimble, 2006) and has a role in self-related mental representations during rest and first-person perspective taking (Vogeley et al., 2001). The PCC, seated at the nexus of multiple intrinsic connectivity networks, including the DMN, has a central role in internally directed cognition (Buckner et al., 2008). The supramarginal gyrus is important for emotion perception. Taken together, altered connectivity between the DMPFC and these regions could manifest clinically as a disturbance of self-appraisal and emotional perception in PPD.

Examining PPD abnormalities within the DMN, an fMRI study scanned women diagnosed with PPD and healthy women at 12 or fewer weeks postpartum (Chase, Moses-Kolko, Zevallos, Wisner, & Phillips, 2014). In women with PPD, depressive symptoms were negatively correlated with PCC and right amygdala connectivity (Chase et al., 2014). This disrupted PCC to right amygdala connectivity may be connected to deficits in mother- infant bonding observed in PPD (Sylven et al., 2013).

In a task-based study of healthy, asymptomatic women 2–3 months postpartum, depressive symptoms, as measured by self-report, were negatively correlated to amygdala activity while viewing photos of their own infants (Barrett et al., 2012). Women 6–8 weeks postpartum experiencing symptoms consistent with a PPD diagnosis showed a significant negative correlation of depressive symptoms and right amygdala activation when viewing threat-related words (i.e., murder), compared to healthy postpartum women (Silverman et al., 2011). Women 4–13 weeks postpartum diagnosed with PPD had reduced left DMPFC and left amygdala activity when viewing negative, adult emotional faces, as compared to healthy postpartum women (Moses-Kolko et al., 2010). Having increased sensitivity to threatening and negative stimuli may be an evolutionary advantage for women to protect their newborns in the postpartum period, but women with PPD may be at disadvantage as they exhibit blunted amygdala activation. As functional studies in nonperipartum depression have shown increased amygdala activation when viewing negative stimuli (Siegle, Steinhauer, Thase, Stenger, & Carter, 2002), these results are critical in further understanding the potential biological differences between PPD and MDD.

In contrast to the findings above, a study that compared healthy postpartum women and women with PPD 2–5 months postpartum found no difference in amygdala BOLD response to viewing photos of their own infant smiling (Wonch et al., 2016). However, women with PPD showed an enhanced, nondiscriminant response in the right amygdala to all other positive stimuli presented (i.e., photos of another infant smiling and noninfant photos), while healthy women showed amygdala activation only in response to their own infants. This may be caused by hypervigilance or even dysregulated arousal for salient stimuli. While these results seem contradictory to the studies that found blunted amygdala activation in PPD, the authors posit that their results are due to the fact that they focused on amygdala response to positive stimuli, rather than negative (Wonch et al., 2016). These results further the notion that PPD may be biologically distinct from MDD, as prior MDD studies have found that amygdala response is decreased in response to positive stimuli (Stuhrmann et al., 2013).

Additionally, higher depressive symptomology was related to decreased right anterior IC and amygdala connectivity, with healthy women showing the opposite (Wonch et al., 2016). The IC is involved in representing one's internal state and emotional awareness (Craig, 2009) and is implicated in anxiety disorders (Paulus & Stein, 2006). Since women with and without PPD scored significantly different on self-reported measures of trait anxiety (Wonch et al., 2016), further exploration of the IC—amygdala relationship may give researchers insight into how anxiety symptoms play a role in phenotypes of PPD. As both the IC and amygdala are located within the same network, these results suggest that PPD may disrupt connectivity within the SN, increasing the need for future research on connectivity between other brain regions within the SN and their relation to PPD symptoms. Collectively the above findings illustrate the critical role the amygdala plays in understanding and identifying functional biomarkers in PPD.

In a study comparing healthy postpartum women and naturally cycling nonpostpartum women, participants completed an emotion face matching task in the scanner. Self-reported, nonclinical depressive symptoms were positively correlated with inferior frontal gyrus and insular reactivity in the late postpartum (Gingnell et al., 2015). Increased insular reactivity may represent an adaptation in emotion processing that increases the risk for depression.

A task-based fMRI study examined reward-related deficits in women diagnosed with PPD and healthy postpartum women at 10 or fewer weeks postpartum (Moses-Kolko et al., 2011). A monetary reward paradigm was used, where women guessed the value of a hidden card and received rewards for correct guesses and losses for incorrect ones. Initially, no differences were observed in response to the reward task, with both healthy women and women with PPD showing positive activity peaks in left ventral striatum activity. However, while the healthy women maintained a response to the reward, the women with PPD attenuated to baseline. Additionally, there was a negative correlation between left striatal activity and depression severity, as measured by self-report, suggesting that striatal activity may play a role in reward-related deficits in PPD (Moses-Kolko et al., 2011).

Overall, studies comparing women with and without PPD point to changes in activation and connectivity in three main networks: the DMN, the SN, and the CEN. Further research is needed on how the disrupted connectivity, within and between the networks, translates to a specific clinical presentation, aiding in detection and potential therapeutic implications. The above findings also indicate that even asymptomatic or subclinical depressive symptoms in women, and not just those consistent with a PPD diagnosis, are associated with differences in connectivity. This is consistent with the clinical literature indicating that both subclinical and clinical nonperipartum depression symptoms are impairing (Felder et al., 2012; Wei et al., 2015). Measuring severity and symptom type will be important in understanding the neurocircuitry of the different presentations of peripartum depression.

Peripartum bipolar depression

Peripartum bipolar depression is underdetected (Jones, Chandra, Dazzan, & Howard, 2014) and often misdiagnosed as unipolar depression (Sharma et al., 2008), which can lead to inadequate treatment and worsening of symptoms. While just 2.8% of peripartum women are diagnosed with bipolar disorder (Vesga-Lopez et al., 2008), the risk of having a first-time inpatient psychiatric admission specifically related to bipolar disorder is 23.33% within the first 30 days of pregnancy (Munk-Olsen, Laursen, Pedersen, Mors, & Mortensen, 2006). Women with a prepregnancy diagnosis of bipolar I disorder have a 50% chance of suffering a relapse of bipolar symptoms during the peripartum period, with more than 20% affected by mania or psychotic depression (Di Florio et al., 2013). Peripartum episodes can cause detrimental effects on mother and child and are linked to substance abuse (Judd et al., 2014), self-injury (Taylor, van Ravesteyn, van denBerg, Stewart, & Howard, 2016), suicide (Khalifeh, Hunt, Appleby, & Howard, 2016), and infanticide (Spinelli, 2004). The diagnosis of bipolar depression can be combined with the peripartum onset specifier if symptoms occur during pregnancy or within 4 weeks following delivery (APA, 2013), yet research in peripartum bipolar disorder is scarce (Sharma, Doobay, & Baczynski, 2017). There is a critical need for better diagnostic tools (Jones et al., 2014; Vesga-Lopez et al., 2008) and further research into the biological underpinnings of peripartum bipolar disorder so that targeted treatments may be developed.

fMRI findings in nonperipartum bipolar disorder

Our literature search did not identify any fMRI studies in peripartum bipolar disorder. To guide future research, we briefly review fMRI findings in individuals with bipolar disorder outside the peripartum period. Individuals with euthymic bipolar I disorder have deficits in functional connectivity between prefrontal cortex and subcortical structures that are not observed in euthymic individuals with bipolar II disorder (Caseras et al., 2015). In bipolar mania, depression, and mixed episodes, it is hypothesized that the amygdala-anterior paralimbic network is disrupted (Blond, Fredericks, & Blumberg, 2012). The ACC integrates dorsal prefrontal (cognitive) and ventral prefrontal (emotional) functions (Strakowski et al., 2011; Yamasaki, LaBar, & McCarthy, 2002) and is also implicated in bipolar I disorder (Altshuler et al., 2005; Strakowski et al., 2011). Finally, individuals with bipolar versus unipolar depression respond differently to emotional stimuli based on ventral and dorsal ACC activity (Burger et al., 2017; Delvecchio et al., 2012).

Peripartum psychosis

Peripartum psychosis occurs in 0.89–2.6 in 1000 peripartum women (VanderKruik et al., 2017). While a prepregnancy diagnosis of bipolar disorder increases the risk of peripartum psychosis to one in five (Di Florio et al., 2013), 50% or more of women who develop postpartum psychosis have no history or risk factors (Blackmore et al., 2013). Peripartum psychosis is associated with increased risk of aggressive behavior, suicide (Lysell et al., 2018), and infanticide (Spinelli, 2004). Women can experience mania, depression (Jones et al., 2014), and obsessive and disorganized thoughts surrounding the infant, such as delusions of self-sacrificing suicide and homicide (Bergink, Rasgon, & Wisner, 2016). Peripartum psychosis is a psychiatric emergency, as sudden onset of psychotic symptoms can cause rapid deterioration. The diagnosis of brief psychotic disorder with the peripartum onset specifier can be used for women who experience the onset of psychosis during pregnancy or within 4 weeks postpartum (APA, 2013). After the first occurrence of postpartum psychosis, there is a 50%–80% chance of experiencing a later psychiatric episode (Nager, Szulkin, Johansson, Johansson, & Sundquist, 2013), yet the underlying pathophysiology is largely unknown.

fMRI findings in nonperipartum psychosis

Our literature search did not identify any fMRI studies in peripartum women with psychotic disorders; however, one structural MRI study was found indicating reductions in volume of the parahippocampal gyrus, ACC, postcentral gyrus, and superior temporal gyrus (STG) in postpartum women who developed a recent psychotic episode compared with at risk women who did not (Fuste et al., 2017). To inform future research, we briefly review fMRI findings in individuals at-risk of psychotic disorders outside the peripartum period. Individuals at clinical high risk (CHR) of developing psychosis, first-episode psychosis, and schizophrenia have similar structural brain abnormalities in the prefrontal, medial temporal, and anterior cingulate regions (Borgwardt et al., 2007; Fornito et al., 2008). A recent quantitative fMRI review examined individuals at CHR of psychosis and identified four main dysfunctional cortical regions: right inferior parietal lobule, left MFG, left STG, and the frontopolar cortex in the right superior frontal gyrus (Dutt et al., 2015). The inferior parietal lobule is involved in working memory (Borst & Anderson, 2013), while activity of the STG has been associated with auditory hallucinations (Allen et al., 2007). It is important to note that while these may be regions of interest to guide initial peripartum psychosis research, it may be the case that peripartum and nonperipartum psychotic disorders present with different neural underpinnings.

Peripartum anxiety disorders and obsessive–compulsive disorder

Peripartum anxiety is diagnosed in 10%–15% of women (Dennis, Falah-Hassani, & Shiri, 2017) and commonly cooccurs with depression (Dindo et al., 2017). Anxiety disorders and OCD during the peripartum period can cause significant adverse effects for the woman and child (Nicol-Harper, Harvey, & Stein, 2007; Stein et al., 2012), with antenatal anxiety the strongest predictor of alcohol use in pregnancy (Arch, 2013). There is no diagnosis or specifier for peripartum OCD or any peripartum anxiety disorder in the DSM-5. Furthermore, there is little research on the course of anxiety disorders in the peripartum period or their pathophysiology. This section will focus on the current functional neuroimaging literature and future directions in peripartum generalized anxiety disorder (GAD), post-traumatic stress disorder (PTSD), and OCD. We chose to review these disorders due to their higher prevalence in peripartum women, comorbidity with depression and psychosis (Beck, Gable, Sakala, & Declercq, 2011; Dindo et al., 2017; Falah-Hassani, Shiri, & Dennis, 2017; Sharma et al., 2008), and availability of information in the literature.

fMRI findings in peripartum anxiety

We identified one fMRI study focusing solely on peripartum anxiety, in women 3–4 months postpartum (Guo, Moses-Kolko, Phillips, Swain, & Hipwell, 2018). At 3 months postpartum, standardized audio recordings were made of their infant's cry. Another standardized audio recording of an unfamiliar infant was also made. At 4 months postpartum, subjects underwent an fMRI infant cry task where they listened to own-infant cry, other-infant cry, and a garbled sound clip before rating their distress and urge to help the baby. Women were then filmed interacting with their infant and rated on warmth, involvement, hostility, intrusiveness, and sensitivity and were assessed for anxiety and depressive symptoms. In women with high anxiety there was a positive association between positive maternal behavior and functional connectivity between the amygdala and right posterior superior temporal sulcus, as well as the amygdala and right STG. There was a 31% increase in positive maternal behavior for each unit increase of functional connectivity between the amygdala and right posterior superior temporal sulcus. In women with low anxiety there was a negative association between positive maternal behavior and functional connectivity between the amygdala and the right posterior superior temporal sulcus. There was a 17% decrease in positive maternal behavior for each unit increase of functional connectivity between the amygdala and right posterior superior temporal sulcus. These findings suggest that the functional connectivity between the amygdala and superior temporal sulcus may distinctly modulate positive maternal behavior in women with high anxiety (Guo et al., 2018).

Peripartum generalized anxiety disorder

While the lifetime prevalence of GAD in the general population is 3.7% (Ruscio et al., 2017), the prevalence of peripartum GAD may be greater, averaging from 8.5% to 10.5% (Buist, Gotman, & Yonkers, 2011) in pregnancy and 4.4%–10.8% (Misri, Abizadeh, Sanders, & Swift, 2015) in postpartum. Women with peripartum GAD can experience worries focused on fears of fetal well-being, maternal health, illness in the partner, and parental death, leading to functional impairment. Postpartum GAD can be associated with less maternal responsiveness and engagement with infants (Stein et al., 2012) with symptoms persisting up to 2 years later (Prenoveau et al., 2013).

fMRI findings in peripartum generalized anxiety disorder

To date, there are no imaging studies that examine GAD in the peripartum period. A systematic review on GAD outside the peripartum period yielded 14 fMRI studies with varying methodologies and results (Mochcovitch, da Rocha Freire, Garcia, & Nardi, 2014). Generalizing these studies is challenging, but a consistent result is that individuals with GAD who perform emotional regulation tasks have alterations in functional connectivity between the amygdala, prefrontal cortex, and ACC (Etkin et al., 2009, 2010).

Peripartum posttraumatic stress disorder

The prevalence of PTSD across the peripartum period averages 4.6%–6.79%, while women at high risk average significantly higher prevalence rates, 16.61%–18.95% (Yildiz, Ayers, & Phillips, 2017). Peripartum PTSD often goes undetected and untreated and can present in a variety of clinical situations (Grekin & O'Hara, 2014). Traumatic childbirth, which can involve unexpected or emergency interventions during labor, can be a risk factor for PTSD development. Women can develop nightmares about labor, thoughts of infant death, reluctance to engage with their infants, or reluctance to bring their infants to the hospitals where they were delivered. Alternatively, women who experienced a prepregnancy traumatic event can continue to experience PTSD symptoms in pregnancy and the postpartum period, or symptoms can be retriggered for women whose PTSD symptoms were previously resolved. Postpartum PTSD symptoms predict poorer child social and emotional development at 2 years postpartum, with the relationship remaining significant when controlling for infant temperament and maternal depression and anxiety (Garthus-Niegel, Ayers, Martini, von Soest, & Eberhard-Gran, 2017).

fMRI findings in peripartum posttraumatic stress disorder

There are no fMRI studies in peripartum women with PTSD within 6 months postpartum. To guide future research, we provide a brief overview of fMRI findings in individuals with PTSD outside the peripartum period. The largest quantitative metaanalysis of resting state fMRI studies of PTSD analyzed 22 studies that included individuals with PTSD, individuals exposed to trauma, and individuals without trauma exposure (Disner, Marquardt, Mueller, Burton, & Sponheim, 2018). Five regions of interest had a significant difference in activity when comparing individuals with and without PTSD: left globus pallidus (all individuals without PTSD), left inferior parietal lobule and right lingual gyrus (trauma exposed individuals), left amygdala, and right caudate head (nontraumatized individuals) (Disner et al., 2018).

Peripartum obsessive–compulsive disorder

The prevalence of OCD in the general population averages 2%–3% (Sasson et al., 1997) and in the peripartum period averages 4%–9% (Goodman, Watson, & Stubbs, 2016), with 11% of women experiencing mild OCD symptoms in the early postpartum period (Miller, Chu, Gollan, & Gossett, 2013). Obsessions and compulsions in the peripartum period can focus on the newborn's health, fear of harming the infant, of the infant being contaminated, and of infant death (Abramowitz, Schwartz, Moore, & Luenzmann, 2003) and have been shown to persist up to 1 year postpartum (Uguz, Akman, Kaya, & Cilli, 2007). While women who experience peripartum psychosis can also experience obsessions and compulsions related to the infant, peripartum OCD is egodystonic (Abramowitz et al., 2003). Some studies report different patterns of OCD symptomology during pregnancy and postpartum (Abramowitz et al., 2003; Uguz et al., 2007), indicating that the time points should be studied separately.

fMRI findings in peripartum obsessive–compulsive disorder

A single fMRI study explored stress reactivity in postpartum women with and without OCD (Lord, Steiner, Soares, Carew, & Hall, 2012). This study recruited 12 women with OCD within 6 months postpartum, 16 healthy postpartum women, and 11 healthy nonpostpartum women for a Montreal Imaging Stress Task while undergoing BOLD fMRI. In women with postpartum OCD, there was significant activation of STG, orbitofrontal cortex, insula, inferior temporal gyri, and medial PFC, whereas healthy postpartum women had significant deactivation of these areas. In healthy postpartum women, there was significant activation in the DLPFC as well as deactivation in the medial PFC and ACC, whereas healthy non-postpartum women had the reverse pattern. This study lends support to the theory of orbitofrontal cortex–STG dysfunction in OCD and elucidates stress-related activation patterns in women with postpartum OCD, post-partum healthy, and nonpostpartum healthy women (Fineberg et al., 2010; Graybiel & Rauch, 2000).

Discussion

The pathophysiology of peripartum psychiatric disorders is poorly understood. While the field is starting to make gains in understanding the neurobiology of unipolar depression, comorbid psychiatric disorders are very common and yet are inconsistently reported in imaging studies. Many PPD imaging studies measure self-reported levels of anxiety, but fail to report comorbid clinical or research interview—derived diagnoses or incorporate them into the understanding of findings. In addition, many studies use the Edinburgh Postnatal Depression Scale (EPDS) (Cox, Holden, & Sagovsky, 1987) to identify women with either symptoms or a "diagnosis" of PPD. A total score of 13 or more, out of a maximum of 30, has been recommended for identifying women with a *possible* diagnosis of major depression (Cox et al., 1987), though studies also use different clinical cutoffs, ranging from 7 to 13 (Chaudron et al., 2010; Matthey, Henshaw, Elliott, & Barnett, 2006). While the EPDS is very helpful in clinical settings to screen quickly for depressive symptoms in peripartum women, it does not assess for all of the symptoms of MDD. The EPDS also detects other major psychiatric illnesses in addition to depression (Lydsdottir et al., 2014). As many studies rely on the EPDS as the sole measure of "depressive symptoms" and sometimes use a cutoff score to denote a "diagnosis of PPD," it is important to recognize that use in this manner is a significant limitation given the evidence above. Future studies should use self-report questionnaires in conjunction with clinical interviews that have clinical applicability in addition to a standardized research interview to accurately diagnose peripartum psychiatric disorders.

While fMRI imaging research is increasing in postpartum women, there are no fMRI studies in pregnancy, when many women first develop symptoms or onset of clinical diagnosis. Studies report different patterns of symptomatology during pregnancy and postpartum, indicating that the time points should be studied separately. The physiological changes of pregnancy and the postpartum period can affect cerebral blood flow and brain connectivity differently. Furthermore, as studies have been conducted specifically on the fetal brain in utero, research has shown that MRI techniques are safe for expecting mothers and their unborn children (Committee Opinion No. 723, 2017; Ray, Vermeulen, Bharatha, Montanera, & Park, 2016). The need for future fMRI studies separating antenatal and postpartum psychiatric illness is paramount in further understanding the etiology and progression of these disorders and their heterogeneity.

Disorders during antepartum and postpartum can have different behavioral phenotypes than the same disorders outside the peripartum period (e.g., unipolar depression). As this may lead to their neural biomarkers also differing, the above reviews outside the peripartum period should be used only as a place to start. With future research, neuroimaging biomarkers will be able to identify women at risk of developing peripartum psychiatric disorders, aid in diagnosing comorbidity, and used to aid in personalized treatment selection. Clearly, we are at an exciting beginning: currently most imaging research aims to distinguish postpartum women with psychiatric illness versus healthy postpartum women, yet future research, as it builds upon this work, will increasingly contribute to clinical care.

Acknowledgment

Janice Lester, MLS, Reference and Education Librarian, Health Science Library, Long Island Jewish Medical Center, Northwell Health.

References

Abramowitz, J. S., Schwartz, S. A., Moore, K. M., & Luenzmann, K. R. (2003). Obsessive-compulsive symptoms in pregnancy and the puerperium: A review of the literature. *Journal of Anxiety Disorders, 17*(4), 461–478.

Allen, P., Amaro, E., Fu, C. H., Williams, S. C., Brammer, M. J., Johns, L. C., & McGuire, P. K. (2007). Neural correlates of the misattribution of speech in schizophrenia. *The British Journal of Psychiatry, 190*, 162–169. https://doi.org/10.1192/bjp.bp.106.025700.

Altshuler, L., Bookheimer, S., Proenza, M. A., Townsend, J., Sabb, F., Firestine, A., … Cohen, M. S. (2005). Increased amygdala activation during mania: A functional magnetic resonance imaging study. *American Journal of Psychiatry, 162*(6), 1211–1213. https://doi.org/10.1176/appi.ajp.162.6.1211.

Anand, A., Li, Y., Wang, Y., Wu, J., Gao, S., Bukhari, L., … Lowe, M. J. (2005). Activity and connectivity of brain mood regulating circuit in depression: A functional magnetic resonance study. *Biological Psychiatry, 57*(10), 1079–1088. https://doi.org/10.1016/j.biopsych.2005.02.021.

APA. (2013). *Diagnostic and statistical manual of mental disorders, DSM-5* (5th ed.). Washington, DC: American Psychiatric Publishing.

Arch, J. J. (2013). Pregnancy-specific anxiety: Which women are highest and what are the alcohol-related risks? *Comprehensive Psychiatry, 54*(3), 217–228. https://doi.org/10.1016/j.comppsych.2012.07.010.

Bannbers, E., Gingnell, M., Engman, J., Morell, A., Sylven, S., Skalkidou, A., … Poromaa, I. S. (2013). Prefrontal activity during response inhibition decreases over time in the postpartum period. *Behavioural Brain Research, 241*, 132–138. https://doi.org/10.1016/j.bbr.2012.12.003.

Barrett, J., & Fleming, A. S. (2011). Annual research review: All mothers are not created equal: Neural and psychobiological perspectives on mothering and the importance of individual differences *Journal of Child Psychology and Psychiatry, 52*(4), 368–397. https://doi.org/10.1111/j.1469-7610.2010.02306.x.

Barrett, J., Wonch, K. E., Gonzalez, A., Ali, N., Steiner, M., Hall, G. B., & Fleming, A. S. (2012). Maternal affect and quality of parenting experiences are related to amygdala response to infant faces. *Social Neuroscience, 7*(3), 252–268. https://doi.org/10.1080/17470919.2011.609907.

Bauer, A., Knapp, M., & Parsonage, M. (2016). Lifetime costs of perinatal anxiety and depression. *Journal of Affective Disorders, 192*, 83–90. https://doi.org/10.1016/j.jad.2015.12.005.

Beck, C. T., Gable, R. K., Sakala, C., & Declercq, E. R. (2011). Posttraumatic stress disorder in new mothers: Results from a two-stage U.S. national survey. *Birth, 38*(3), 216–227. https://doi.org/10.1111/j.1523-536X.2011.00475.x.

Bergink, V., Rasgon, N., & Wisner, K. L. (2016). Postpartum psychosis: madness, mania, and melancholia in motherhood. *American Journal of Psychiatry, 173*(12), 1179–1188. https://doi.org/10.1176/appi.ajp.2016.16040454.

Biswal, B., Yetkin, F. Z., Haughton, V. M., & Hyde, J. S. (1995). Functional connectivity in the motor cortex of resting human brain using echo-planar MRI. *Magnetic Resonance in Medicine, 34*(4), 537–541.

Blackmore, E. R., Rubinow, D. R., O'Connor, T. G., Liu, X., Tang, W., Craddock, N., & Jones, I. (2013). Reproductive outcomes and risk of subsequent illness in women diagnosed with postpartum psychosis. *Bipolar Disorders, 15*(4), 394–404. https://doi.org/10.1111/bdi.12071.

Blond, B. N., Fredericks, C. A., & Blumberg, H. P. (2012). Functional neuroanatomy of bipolar disorder: Structure, function, and connectivity in an amygdala-anterior paralimbic neural system. *Bipolar Disorders, 14*(4), 340–355. https://doi.org/10.1111/j.1399-5618.2012.01015.x.

Borgwardt, S. J., Riecher-Rossler, A., Dazzan, P., Chitnis, X., Aston, J., Drewe, M., … McGuire, P. K. (2007). Regional gray matter volume abnormalities in the at risk mental state. *Biological Psychiatry, 61*(10), 1148–1156. https://doi.org/10.1016/j.biopsych.2006.08.009.

Borst, J. P., & Anderson, J. R. (2013). Using model-based functional MRI to locate working memory updates and declarative memory retrievals in the fronto-parietal network. *Proceedings of the National Academy of Sciences of the United States of America, 110*(5), 1628–1633. https://doi.org/10.1073/pnas.1221572110.

Bressler, S. L., & Menon, V. (2010). Large-scale brain networks in cognition: Emerging methods and principles. *Trends in Cognitive Sciences, 14*(6), 277–290. https://doi.org/10.1016/j.tics.2010.04.004.

Buckner, R. L., Andrews-Hanna, J. R., & Schacter, D. L. (2008). The brain's default network: Anatomy, function, and relevance to disease. *Annals of the New York Academy of Sciences, 1124*, 1–38. https://doi.org/10.1196/annals.1440.011.

Buckwalter, J. G., Stanczyk, F. Z., McCleary, C. A., Bluestein, B. W., Buckwalter, D. K., Rankin, K. P., … Goodwin, T. M. (1999). Pregnancy, the postpartum, and steroid hormones: Effects on cognition and mood. *Psychoneuroendocrinology, 24*(1), 69–84.

Buist, A., Gotman, N., & Yonkers, K. A. (2011). Generalized anxiety disorder: Course and risk factors in pregnancy. *Journal of Affective Disorders, 131*(1–3), 277–283. https://doi.org/10.1016/j.jad.2011.01.003.

Burger, C., Redlich, R., Grotegerd, D., Meinert, S., Dohm, K., Schneider, I., … Dannlowski, U. (2017). Differential abnormal pattern of anterior cingulate gyrus activation in unipolar and bipolar depression: An fMRI and pattern classification approach. *Neuropsychopharmacology, 42*(7), 1399–1408. https://doi.org/10.1038/npp.2017.36.

Byatt, N., Carter, D., Deligiannidis, K. M., Epperson, C. N., Meltzer-Brody, S., Payne, J. L., … Yonkers, K. (2018). *Position statement on screening and treatment of mood and anxiety disorders during pregnancy and postpartum*. American Psychiatric Association.

Caseras, X., Murphy, K., Lawrence, N. S., Fuentes-Claramonte, P., Watts, J., Jones, D. K., & Phillips, M. L. (2015). Emotion regulation deficits in euthymic bipolar I versus bipolar II disorder: A functional and diffusion-tensor imaging study. *Bipolar Disorders, 17*(5), 461–470. https://doi.org/10.1111/bdi.12292.

Cavanna, A. E., & Trimble, M. R. (2006). The precuneus: A review of its functional anatomy and behavioural correlates. *Brain, 129*(Pt 3), 564–583. https://doi.org/10.1093/brain/awl004.

Chase, H. W., Moses-Kolko, E. L., Zevallos, C., Wisner, K. L., & Phillips, M. L. (2014). Disrupted posterior cingulate-amygdala connectivity in postpartum depressed women as measured with resting BOLD fMRI. *Social Cognitive and Affective Neuroscience, 9*(8), 1069–1075. https://doi.org/10.1093/scan/nst083.

Chaudron, L. H., Szilagyi, P. G., Tang, W., Anson, E., Talbot, N. L., Wadkins, H. I., ... Wisner, K. L. (2010). Accuracy of depression screening tools for identifying postpartum depression among urban mothers. *Pediatrics, 125*(3), e609–617. https://doi.org/10.1542/peds.2008-3261.

Committee Opinion No 723. (2017). Guidelines for diagnostic imaging during pregnancy and lactation. *Obstetrics & Gynecology, 130*(4), e210–e216. https://doi.org/10.1097/aog.0000000000002355.

Cox, J. L., Holden, J. M., & Sagovsky, R. (1987). Detection of postnatal depression. Development of the 10-item Edinburgh postnatal depression scale. *The British Journal of Psychiatry, 150*, 782–786.

Craig, A. D. (2009). How do you feel—now? The anterior insula and human awareness. *Nature Reviews Neuroscience, 10*(1), 59–70. https://doi.org/10.1038/nrn2555.

Deligiannidis, K. M., Fales, C. L., Kroll-Desrosiers, A. R., Shaffer, S. A., Villamarin, V., Tan, Y., ... Moore, C. M. (2019). Resting-state functional connectivity, cortical GABA, and neuroactive steroids in peripartum and peripartum depressed women: A functional magnetic resonance imaging and spectroscopy study. *Neuropsychopharmacology, 44*(3), 546–554. https://doi.org/10.1038/s41386-018-0242-2.

Deligiannidis, K. M., Sikoglu, E. M., Shaffer, S. A., Frederick, B., Svenson, A. E., Kopoyan, A., ... Moore, C. M. (2013). GABAergic neuroactive steroids and resting-state functional connectivity in postpartum depression: A preliminary study. *Journal of Psychiatric Research, 47*(6), 816–828. https://doi.org/10.1016/j.jpsychires.2013.02.010.

Delvecchio, G., Fossati, P., Boyer, P., Brambilla, P., Falkai, P., Gruber, O., ... Frangou, S. (2012). Common and distinct neural correlates of emotional processing in bipolar disorder and major depressive disorder: A voxel-based meta-analysis of functional magnetic resonance imaging studies. *European Neuropsychopharmacology, 22*(2), 100–113. https://doi.org/10.1016/j.euroneuro.2011.07.003.

Dennis, C. L., Falah-Hassani, K., & Shiri, R. (2017). Prevalence of antenatal and postnatal anxiety: Systematic review and meta-analysis. *The British Journal of Psychiatry, 210*(5), 315–323. https://doi.org/10.1192/bjp.bp.116.187179.

Di Florio, A., Forty, L., Gordon-Smith, K., Heron, J., Jones, L., Craddock, N., & Jones, I. (2013). Perinatal episodes across the mood disorder spectrum. *JAMA Psychiatry, 70*(2), 168–175. https://doi.org/10.1001/jamapsychiatry.2013.279.

Dindo, L., Elmore, A., O'Hara, M., & Stuart, S. (2017). The comorbidity of axis I disorders in depressed pregnant women. *Archives of Women's Mental Health, 20*(6), 757–764. https://doi.org/10.1007/s00737-017-0769-y.

Disner, S. G., Marquardt, C. A., Mueller, B. A., Burton, P. C., & Sponheim, S. R. (2018). Spontaneous neural activity differences in posttraumatic stress disorder: A quantitative resting-state meta-analysis and fMRI validation. *Human Brain Mapping, 39*(2), 837–850. https://doi.org/10.1002/hbm.23886

Draper, E. S., Gallimore, I. D., Kurinczuk, J. J., Smith, P. W., Boby, T., Smith, L. K., , ... Manktelow, B. N., & on behalf of the MBRRACE-UK Collaboration. (2018). *MBRRACE-UK perinatal mortality surveillance report, UK perinatal deaths for births from january to december 2016.* University of Leicester, UK: The Infant Mortality and Morbidity Studies.

Dutt, A., Tseng, H. H., Fonville, L., Drakesmith, M., Su, L., Evans, J., ... David, A. S. (2015). Exploring neural dysfunction in 'clinical high risk' for psychosis: A quantitative review of fMRI studies. *Journal of Psychiatric Research, 61*, 122–134. https://doi.org/10.1016/j.jpsychires.2014.08.018.

Etkin, A., Prater, K. E., Hoeft, F., Menon, V., & Schatzberg, A. F. (2010). Failure of anterior cingulate activation and connectivity with the amygdala during implicit regulation of emotional processing in generalized anxiety disorder. *American Journal of Psychiatry, 167*(5), 545–554. https://doi.org/10.1176/appi.ajp.2009.09070931.

Etkin, A., Prater, K. E., Schatzberg, A. F., Menon, V., & Greicius, M. D. (2009). Disrupted amygdalar subregion functional connectivity and evidence of a compensatory network in generalized anxiety disorder. *Archives of General Psychiatry, 66*(12), 1361–1372. https://doi.org/10.1001/archgenpsychiatry.2009.104.

Falah-Hassani, K., Shiri, R., & Dennis, C. L. (2017). The prevalence of antenatal and postnatal co-morbid anxiety and depression: A meta-analysis. *Psychological Medicine, 47*(12), 2041–2053. https://doi.org/10.1017/S0033291717000617.

Felder, J. N., Smoski, M. J., Kozink, R. V., Froeliger, B., McClernon, J., Bizzell, J., ... Dichter, G. S. (2012). Neural mechanisms of subclinical depressive symptoms in women: A pilot functional brain imaging study. *BMC Psychiatry, 12*, 152. https://doi.org/10.1186/1471-244X-12-152.

Fineberg, N. A., Potenza, M. N., Chamberlain, S. R., Berlin, H. A., Menzies, L., Bechara, A., ... Hollander, E. (2010). Probing compulsive and impulsive behaviors, from animal models to endophenotypes: A narrative review. *Neuropsychopharmacology, 35*(3), 591–604. https://doi.org/10.1038/npp.2009.185.

Fisher, J., Cabral de Mello, M., Patel, V., Rahman, A., Tran, T., Holton, S., & Holmes, W. (2012). Prevalence and determinants of common perinatal mental disorders in women in low- and lower-middle-income countries: A systematic review. *Bulletin of the World Health Organization, 90*(2), 139G–149G. https://doi.org/10.2471/BLT.11.091850.

Fornito, A., Yung, A. R., Wood, S. J., Phillips, L. J., Nelson, B., Cotton, S., ... Yucel, M. (2008). Anatomic abnormalities of the anterior cingulate cortex before psychosis onset: An MRI study of ultra-high-risk individuals. *Biological Psychiatry, 64*(9), 758–765. https://doi.org/10.1016/j.biopsych.2008.05.032.

Fuste, M., Pauls, A., Worker, A., Reinders, A., Simmons, A., Williams, S. C. R., ... Dazzan, P. (2017). Brain structure in women at risk of postpartum psychosis: An MRI study. *Translational Psychiatry,* 7(12), 1286. https://doi.org/10.1038/s41398-017-0003-8.

Garthus-Niegel, S., Ayers, S., Martini, J., von Soest, T., & Eberhard-Gran, M. (2017). The impact of postpartum post-traumatic stress disorder symptoms on child development: A population-based, 2-year follow-up study. *Psychological Medicine, 47*(1), 161–170. https://doi.org/10.1017/S003329171600235X.

Gingnell, M., Bannbers, E., Moes, H., Engman, J., Sylven, S., Skalkidou, A., ... Sundstrom-Poromaa, I. (2015). Emotion reactivity is increased 4-6 weeks postpartum in healthy women: A longitudinal fMRI study. *PLoS One, 10*(6), e0128964. https://doi.org/10.1371/journal.pone.0128964.

Goodman, J. H., Watson, G. R., & Stubbs, B. (2016). Anxiety disorders in postpartum women: A systematic review and meta-analysis. *Journal of Affective Disorders, 203*, 292–331. https://doi.org/10.1016/j.jad.2016.05.033.

Goodman, S. H., Rouse, M. H., Connell, A. M., Broth, M. R., Hall, C. M., & Heyward, D. (2011). Maternal depression and child psychopathology: A meta-analytic review. *Clinical Child and Family Psychology Review, 14*(1), 1–27. https://doi.org/10.1007/s10567-010-0080-1.

Graybiel, A. M., & Rauch, S. L. (2000). Toward a neurobiology of obsessive-compulsive disorder. *Neuron, 28*(2), 343–347.

Grekin, R., & O'Hara, M. W. (2014). Prevalence and risk factors of postpartum post-traumatic stress disorder: A meta-analysis. *Clinical Psychology Review, 34*(5), 389–401. https://doi.org/10.1016/j.cpr.2014.05.003.

Grote, N. K., Bridge, J. A., Gavin, A. R., Melville, J. L., Iyengar, S., & Katon, W. J. (2010). A meta-analysis of depression during pregnancy and the risk of preterm birth, low birth weight, and intrauterine growth restriction. *Archives of General Psychiatry, 67*(10), 1012–1024. https://doi.org/10.1001/archgenpsychiatry.2010.111.

Guo, C., Moses-Kolko, E., Phillips, M., Swain, J. E., & Hipwell, A. E. (2018). Severity of anxiety moderates the association between neural circuits and maternal behaviors in the postpartum period. *Cognitive, Affective, & Behavioral Neuroscience, 18*(3), 426–436. https://doi.org/10.3758/s13415-017-0516-x.

Hariri, A. R., Mattay, V. S., Tessitore, A., Kolachana, B., Fera, F., Goldman, D., … Weinberger, D. R. (2002). Serotonin transporter genetic variation and the response of the human amygdala. *Science, 297*(5580), 400–403. https://doi.org/10.1126/science.1071829.

Hendrick, V., Altshuler, L., Strouse, T., & Grosser, S. (2000). Postpartum and non-postpartum depression: Differences in presentation and response to pharmacologic treatment. *Depression and Anxiety, 11*(2), 66–72.

Henry, J. F., & Sherwin, B. B. (2012). Hormones and cognitive functioning during late pregnancy and postpartum: A longitudinal study. *Behavioral Neuroscience, 126*(1), 73–85. https://doi.org/10.1037/a0025540.

Hirschfeld, R. M. (2001). The comorbidity of major depression and anxiety disorders: Recognition and management in primary care. *Primary Care Companion to the Journal of Clinical Psychiatry, 3*(6), 244–254.

Hung, C. I., Liu, C. Y., & Yang, C. H. (2017). Untreated duration predicted the severity of depression at the two-year follow-up point. *PLoS One, 12*(9), e0185119. https://doi.org/10.1371/journal.pone.0185119.

Jones, I., Chandra, P. S., Dazzan, P., & Howard, L. M. (2014). Bipolar disorder, affective psychosis, and schizophrenia in pregnancy and the post-partum period. *Lancet, 384*(9956), 1789–1799. https://doi.org/10.1016/S0140-6736(14)61278-2.

Jones, I., Hamshere, M., Nangle, J. M., Bennett, P., Green, E., Heron, J., … Craddock, N. (2007). Bipolar affective puerperal psychosis: Genome wide significant evidence for linkage to chromosome 16. *American Journal of Psychiatry, 164*(7), 1099–1104. https://doi.org/10.1176/ajp.2007.164.7.1099.

Judd, F., Komiti, A., Sheehan, P., Newman, L., Castle, D., & Everall, I. (2014). Adverse obstetric and neonatal outcomes in women with severe mental illness: To what extent can they be prevented? *Schizophrenia Research, 157*(1–3), 305–309. https://doi.org/10.1016/j.schres.2014.05.030.

Kaiser, R. H., Andrews-Hanna, J. R., Wager, T. D., & Pizzagalli, D. A. (2015). Large-scale network dysfunction in major depressive disorder: A meta-analysis of resting-state functional connectivity. *JAMA Psychiatry, 72*(6), 603–611. https://doi.org/10.1001/jamapsychiatry.2015.0071.

Khalifeh, H., Hunt, I. M., Appleby, L., & Howard, L. M. (2016). Suicide in perinatal and non-perinatal women in contact with psychiatric services: 15 year findings from a UK national inquiry. *The Lancet Psychiatry, 3*(3), 233–242. https://doi.org/10.1016/s2215-0366(16)00003-1.

Kingston, D., Kehler, H., Austin, M. P., Mughal, M. K., Wajid, A., Vermeyden, L., … Giallo, R. (2018). Trajectories of maternal depressive symptoms during pregnancy and the first 12 months postpartum and child externalizing and internalizing behavior at three years. *PLoS One, 13*(4), e0195365. https://doi.org/10.1371/journal.pone.0195365.

Kroll-Desrosiers, A. R., Nephew, B. C., Babb, J. A., Guilarte-Walker, Y., Moore Simas, T. A., & Deligiannidis, K. M. (2017). Association of peripartum synthetic oxytocin administration and depressive and anxiety disorders within the first postpartum year. *Depression and Anxiety, 34*(2), 137–146. https://doi.org/10.1002/da.22599.

Lindahl, V., Pearson, J. L., & Colpe, L. (2005). Prevalence of suicidality during pregnancy and the postpartum. *Archives of Women's Mental Health, 8*(2), 77–87. https://doi.org/10.1007/s00737-005-0080-1.

Logothetis, N. K., Pauls, J., Augath, M., Trinath, T., & Oeltermann, A. (2001). Neurophysiological investigation of the basis of the fMRI signal. *Nature, 412*(6843), 150–157. https://doi.org/10.1038/35084005.

Lord, C., Steiner, M., Soares, C. N., Carew, C. L., & Hall, G. B. (2012). Stress response in postpartum women with and without obsessive-compulsive symptoms: An fMRI study. *Journal of Psychiatry & Neuroscience, 37*(2), 78–86. https://doi.org/10.1503/jpn.110005.

Lydsdottir, L. B., Howard, L. M., Olafsdottir, H., Thome, M., Tyrfingsson, P., & Sigurdsson, J. F. (2014). The mental health characteristics of pregnant women with depressive symptoms identified by the Edinburgh postnatal depression scale. *Journal of Clinical Psychiatry, 75*(4), 393–398. https://doi.org/10.4088/JCP.13m08646.

Lysell, H., Dahlin, M., Viktorin, A., Ljungberg, E., D'Onofrio, B. M., Dickman, P., & Runeson, B. (2018). Maternal suicide - register based study of all suicides occurring after delivery in Sweden 1974–2009. *PLoS One, 13*(1), e0190133. https://doi.org/10.1371/journal.pone.0190133.

Magiakou, M. A., Mastorakos, G., Rabin, D., Dubbert, B., Gold, P. W., & Chrousos, G. P. (1996). Hypothalamic corticotropin-releasing hormone suppression during the postpartum period: Implications for the increase in psychiatric manifestations at this time. *The Journal of Cinical Endocrinology and Metabolism, 81*(5), 1912–1917.

Matthey, S., Henshaw, C., Elliott, S., & Barnett, B. (2006). Variability in use of cut-off scores and formats on the Edinburgh postnatal depression scale: Implications for clinical and research practice. *Archives of Women's Mental Health, 9*(6), 309–315. https://doi.org/10.1007/s00737-006-0152-x.

Miller, E. S., Chu, C., Gollan, J., & Gossett, D. R. (2013). Obsessive-compulsive symptoms during the postpartum period. A prospective cohort. *Journal of Reproductive Medicine, 58*(3–4), 115–122.

Misri, S., Abizadeh, J., Sanders, S., & Swift, E. (2015). Perinatal generalized anxiety disorder: Assessment and treatment. *Journal of Women's Health, 24*(9), 762–770. https://doi.org/10.1089/jwh.2014.5150.

Mochcovitch, M. D., da Rocha Freire, R. C., Garcia, R. F., & Nardi, A. E. (2014). A systematic review of fMRI studies in generalized anxiety disorder: Evaluating its neural and cognitive basis. *Journal of Affective Disorders, 167*, 336–342. https://doi.org/10.1016/j.jad.2014.06.041.

Moses-Kolko, E. L., Fraser, D., Wisner, K. L., James, J. A., Saul, A. T., Fiez, J. A., & Phillips, M. L. (2011). Rapid habituation of ventral striatal response to reward receipt in postpartum depression. *Biological Psychiatry, 70*(4), 395–399. https://doi.org/10.1016/j.biopsych.2011.02.021.

Moses-Kolko, E. L., Perlman, S. B., Wisner, K. L., James, J., Saul, A. T., & Phillips, M. L. (2010). Abnormally reduced dorsomedial prefrontal cortical activity and effective connectivity with amygdala in response to negative emotional faces in postpartum depression. *American Journal of Psychiatry, 167*(11), 1373–1380. https://doi.org/10.1176/appi.ajp.2010.09081235.

Munk-Olsen, T., Laursen, T. M., Pedersen, C. B., Mors, O., & Mortensen, P. B. (2006). New parents and mental disorders: A population-based register study. *Journal of the American Medical Association, 296*(21), 2582–2589. https://doi.org/10.1001/jama.296.21.2582.

Nager, A., Szulkin, R., Johansson, S. E., Johansson, L. M., & Sundquist, K. (2013). High lifelong relapse rate of psychiatric disorders among women with postpartum psychosis. *Nordic Journal of Psychiatry, 67*(1), 53–58. https://doi.org/10.3109/08039488.2012.675590.

Newport, D. J., Levey, L. C., Pennell, P. B., Ragan, K., & Stowe, Z. N. (2007). Suicidal ideation in pregnancy: Assessment and clinical implications. *Archives of Women's Mental Health, 10*(5), 181–187. https://doi.org/10.1007/s00737-007-0192-x.

Nicol-Harper, R., Harvey, A. G., & Stein, A. (2007). Interactions between mothers and infants: Impact of maternal anxiety. *Infant Behavior and Development, 30*(1), 161–167. https://doi.org/10.1016/j.infbeh.2006.08.005.

O'Connor, T. G., Heron, J., Golding, J., Beveridge, M., & Glover, V. (2002). Maternal antenatal anxiety and children's behavioural/emotional problems at 4 years. Report from the avon longitudinal study of parents and children. *The British Journal of Psychiatry, 180*, 502–508.

Parry, B. L., Meliska, C. J., Sorenson, D. L., Lopez, A. M., Martinez, L. F., Nowakowski, S., … Kripke, D. F. (2008). Plasma melatonin circadian rhythm disturbances during pregnancy and postpartum in depressed women and women with personal or family histories of depression. *American Journal of Psychiatry, 165*(12), 1551–1558. https://doi.org/10.1176/appi.ajp.2008.08050709.

Paulus, M. P., & Stein, M. B. (2006). An insular view of anxiety. *Biological Psychiatry, 60*(4), 383–387. https://doi.org/10.1016/j.biopsych.2006.03.042.

Plant, D. T., Pariante, C. M., Sharp, D., & Pawlby, S. (2015). Maternal depression during pregnancy and offspring depression in adulthood: Role of child maltreatment. *The British Journal of Psychiatry, 207*(3), 213–220. https://doi.org/10.1192/bjp.bp.114.156620.

Prenoveau, J., Craske, M., Counsell, N., West, V., Davies, B., Cooper, P., … Stein, A. (2013). Postpartum GAD is a risk factor for postpartum MDD: The course and longitudinal relationships of postpartum GAD and MDD. *Depression and Anxiety, 30*(6), 506–514. https://doi.org/10.1002/da.22040.

Putnam, K., Robertson-Blackmore, E., Sharkey, K., Payne, J., Bergink, V., Munk-Olsen, T., … Meltzer-Brody, S. (2015). Heterogeneity of postpartum depression: A latent class analysis. *The Lancet Psychiatry, 2*(1), 59–67. https://doi.org/10.1016/s2215-0366(14)00055-8.

Putnam, K. T., Wilcox, M., Robertson-Blackmore, E., Sharkey, K., Bergink, V., Munk-Olsen, T., … Treatment, Consortium. (2017). Clinical phenotypes of perinatal depression and time of symptom onset: Analysis of data from an international consortium. *The Lancet Psychiatry, 4*(6), 477–485. https://doi.org/10.1016/S2215-0366(17)30136-0.

Ray, J. G., Vermeulen, M. J., Bharatha, A., Montanera, W. J., & Park, A. L. (2016). Association between MRI exposure during pregnancy and fetal and childhood outcomes. *Journal of the American Medical Association, 316*(9), 952–961. https://doi.org/10.1001/jama.2016.12126.

Ruscio, A. M., Hallion, L. S., Lim, C. C. W., Aguilar-Gaxiola, S., Al-Hamzawi, A., Alonso, J., … Scott, K. M. (2017). Cross-sectional comparison of the epidemiology of DSM-5 generalized anxiety disorder across the globe. *JAMA Psychiatry, 74*(5), 465–475. https://doi.org/10.1001/jamapsychiatry.2017.0056.

Sasson, Y., Zohar, J., Chopra, M., Lustig, M., Iancu, I., & Hendler, T. (1997). Epidemiology of obsessive-compulsive disorder: A world view. *Journal of Clinical Psychiatry, 58*(Suppl. 12), 7–10.

Seeley, W. W., Menon, V., Schatzberg, A. F., Keller, J., Glover, G. H., Kenna, H., … Greicius, M. D. (2007). Dissociable intrinsic connectivity networks for salience processing and executive control. *Journal of Neuroscience, 27*(9), 2349–2356. https://doi.org/10.1523/jneurosci.5587-06.2007.

Sharma, V., Doobay, M., & Baczynski, C. (2017). Bipolar postpartum depression: An update and recommendations. *Journal of affective disorders, 219*, 105–111.

Sharma, V., Khan, M., Corpse, C., & Sharma, P. (2008). Missed bipolarity and psychiatric comorbidity in women with postpartum depression. *Bipolar Disorders, 10*(6), 742–747. https://doi.org/10.1111/j.1399-5618.2008.00606.x.

Siegle, G. J., Steinhauer, S. R., Thase, M. E., Stenger, V. A., & Carter, C. S. (2002). Can't shake that feeling: Event-related fMRI assessment of sustained amygdala activity in response to emotional information in depressed individuals. *Biological Psychiatry, 51*(9), 693–707.

Silverman, M. E., Loudon, H., Liu, X., Mauro, C., Leiter, G., & Goldstein, M. A. (2011). The neural processing of negative emotion postpartum: A preliminary study of amygdala function in postpartum depression. *Archives of Women's Mental Health, 14*(4), 355–359. https://doi.org/10.1007/s00737-011-0226-2.

Smitha, K. A., Arun, K. M., Rajesh, P. G., Thomas, B., & Kesavadas, C. (2017). Resting-state seed-based analysis: An alternative to task-based language fMRI and its laterality index. *American Journal of Neuroradiology, 38*(6), 1187–1192. https://doi.org/10.3174/ajnr.A5169.

Spinelli, M. G. (2004). Maternal infanticide associated with mental illness: Prevention and the promise of saved lives. *American Journal of Psychiatry, 161*(9), 1548–1557. https://doi.org/10.1176/appi.ajp.161.9.1548.

Sridharan, D., Levitin, D. J., & Menon, V. (2008). A critical role for the right fronto-insular cortex in switching between central-executive and default-mode networks. *Proceedings of the National Academy of Sciences of the United States of America, 105*(34), 12569–12574. https://doi.org/10.1073/pnas.0800005105.

Stein, A., Craske, M. G., Lehtonen, A., Harvey, A., Savage-McGlynn, E., Davies, B., … Counsell, N. (2012). Maternal cognitions and mother-infant interaction in postnatal depression and generalized anxiety disorder. *Journal of Abnormal Psychology, 121*(4), 795–809. https://doi.org/10.1037/a0026847.

Strakowski, S. M., Eliassen, J. C., Lamy, M., Cerullo, M. A., Allendorfer, J. B., Madore, M., … Adler, C. M. (2011). Functional magnetic resonance imaging brain activation in bipolar mania: Evidence for disruption of the ventrolateral prefrontal-amygdala emotional pathway. *Biological Psychiatry, 69*(4), 381–388. https://doi.org/10.1016/j.biopsych.2010.09.019.

Stuhrmann, A., Dohm, K., Kugel, H., Zwanzger, P., Redlich, R., Grotegerd, D., … Dannlowski, U. (2013). Mood-congruent amygdala responses to subliminally presented facial expressions in major depression: Associations with anhedonia. *Journal of Psychiatry & Neuroscience, 38*(4), 249–258. https://doi.org/10.1503/jpn.120060.

Sylven, S. M., Elenis, E., Michelakos, T., Larsson, A., Olovsson, M., Poromaa, I. S., & Skalkidou, A. (2013). Thyroid function tests at delivery and risk for postpartum depressive symptoms. *Psychoneuroendocrinology, 38*(7), 1007–1013. https://doi.org/10.1016/j.psyneuen.2012.10.004.

Taylor, C. L., van Ravesteyn, L. M., van denBerg, M. P., Stewart, R. J., & Howard, L. M. (2016). The prevalence and correlates of self-harm in pregnant women with psychotic disorder and bipolar disorder. *Archives of Women's Mental Health, 19*(5), 909–915. https://doi.org/10.1007/s00737-016-0636-2.

Uguz, F., Akman, C., Kaya, N., & Cilli, A. S. (2007). Postpartum-onset obsessive-compulsive disorder: Incidence, clinical features, and related factors. *Journal of Clinical Psychiatry, 68*(1), 132–138.

van den Heuvel, M. P., & Hulshoff Pol, H. E. (2010). Exploring the brain network: A review on resting-state fMRI functional connectivity. *European Neuropsychopharmacology, 20*(8), 519–534. https://doi.org/10.1016/j.euroneuro.2010.03.008.

VanderKruik, R., Barreix, M., Chou, D., Allen, T., Say, L., Cohen, L. S., & Maternal Morbidity Working Group. (2017). The global prevalence of postpartum psychosis: A systematic review. *BMC Psychiatry, 17*(1), 272. https://doi.org/10.1186/s12888-017-1427-7.

Vesga-Lopez, O., Blanco, C., Keyes, K., Olfson, M., Grant, B. F., & Hasin, D. S. (2008). Psychiatric disorders in pregnant and postpartum women in the United States. *Archives of General Psychiatry, 65*(7), 805–815. https://doi.org/10.1001/archpsyc.65.7.805.

Vogeley, K., Bussfeld, P., Newen, A., Herrmann, S., Happe, F., Falkai, P., … Zilles, K. (2001). Mind reading: Neural mechanisms of theory of mind and self-perspective. *NeuroImage, 14*(1 Pt 1), 170–181. https://doi.org/10.1006/nimg.2001.0789.

Wei, X., Shen, H., Ren, J., Liu, W., Yang, R., Liu, J., … Jiang, X. (2015). Alteration of spontaneous neuronal activity in young adults with non-clinical depressive symptoms. *Psychiatry Research, 233*(1), 36–42. https://doi.org/10.1016/j.pscychresns.2015.04.008.

Wisner, K. L., Sit, D. K., McShea, M. C., Rizzo, D. M., Zoretich, R. A., Hughes, C. L., … Hanusa, B. H. (2013). Onset timing, thoughts of self-harm, and diagnoses in postpartum women with screen-positive depression findings. *JAMA Psychiatry, 70*(5), 490–498. https://doi.org/10.1001/jamapsychiatry.2013.87.

Wonch, K. E., de Medeiros, C. B., Barrett, J. A., Dudin, A., Cunningham, W. A., Hall, G. B., … Fleming, A. S. (2016). Postpartum depression and brain response to infants: Differential amygdala response and connectivity. *Social Neuroscience, 11*(6), 600–617. https://doi.org/10.1080/17470919.2015.1131193.

Woody, C. A., Ferrari, A. J., Siskind, D. J., Whiteford, H. A., & Harris, M. G. (2017). A systematic review and meta-regression of the prevalence and incidence of perinatal depression. *Journal of Affective Disorders, 219*, 86–92. https://doi.org/10.1016/j.jad.2017.05.003.

Yamasaki, H., LaBar, K. S., & McCarthy, G. (2002). Dissociable prefrontal brain systems for attention and emotion. *Proceedings of the National Academy of Sciences of the United States of America, 99*(17), 11447–11451. https://doi.org/10.1073/pnas.182176499.

Yildiz, P. D., Ayers, S., & Phillips, L. (2017). The prevalence of posttraumatic stress disorder in pregnancy and after birth: A systematic review and meta-analysis. *Journal of Affective Disorders, 208*, 634–645. https://doi.org/10.1016/j.jad.2016.10.009.

Zheng, J. X., Chen, Y. C., Chen, H., Jiang, L., Bo, F., Feng, Y., … Gu, J. P. (2018). Disrupted spontaneous neural activity related to cognitive impairment in postpartum women. *Frontiers in Psychology, 9*, 624. https://doi.org/10.3389/fpsyg.2018.00624.

CHAPTER 14

Synthesis and future directions

Lauren M. Osborne[1,2,3], **Jennifer L. Payne**[1,2,3]
[1]Department of Psychiatry & Behavioral Sciences, The Johns Hopkins University School of Medicine, Baltimore, MD, United States of America; [2]Department of Gynecology & Obstetrics, The Johns Hopkins University School of Medicine, Baltimore, MD, United States of America; [3]Women's Mood Disorders Center, The Johns Hopkins University School of Medicine, Baltimore, MD, United States of America

We began this project with the idea that this book would serve not only as a roadmap for where research on postpartum mood and anxiety disorders has been but also as a guide for where research in this critical area needs to go. After reading and editing all the chapters in this volume, we realized that a number of themes have emerged.

Because the rate of psychiatric illness increases in the immediate postpartum time period, this period represents a unique research opportunity to identify the biological causes of psychiatric illness more generally

Postpartum-onset psychiatric illness offers a lens through which to observe and measure biological and environmental causes as well as biomarkers of psychiatric illness more generally. Because we cannot predict the onset of bipolar disorder or obsessive compulsive disorder at any other point in life, it is difficult to follow subjects prospectively and identify precedents to the development of these illnesses. Yet in the postpartum period we *can* predict onset—in a sample of 100 women from the general population (which includes both women with and without a prior history of mood disorders), we would expect that approximately 15 of them would develop postpartum depression (Campbell & Cohn, 1991; Dietz et al., 2007; O'Hara, 2009; Yonkers et al., 2001) and 2–3 would develop postpartum OCD (Russell, Fawcett, & Mazmanian, 2013). If you take a sample of 100 women with preexisting psychiatric disorders, these rates skyrocket; 30%–60% of women with preexisting mood disorders will develop postpartum depression (Bratfos & Haug, 1966; Cox, Holden, & Sagovsky, 1987; Davidson & Robertson, 1985; Kendell, Chalmers, & Platz, 1987; Payne et al., 2007; Reich & Winokur, 1970; Viguera et al., 2011; Wisner et al., 2001, 2004), and 20%–30% of those with preexisting bipolar I

Biomarkers of Postpartum Psychiatric Disorders
ISBN 978-0-12-815508-0
https://doi.org/10.1016/B978-0-12-815508-0.00014-X

disorder will develop postpartum psychosis (Wesseloo et al., 2016). Multisite studies, such as the one the PACT consortium is attempting (see Chapter 1), will enlarge sample sizes even further and allow more sophisticated analyses of the underlying biological basis for psychiatric illness in this critical time period. If we are lucky, we will be able to extrapolate findings to psychiatric illness in general, which may, in turn, lead to improved treatments and outcomes.

There is much more research on postpartum depression than other postpartum psychiatric illnesses

It is striking how many more research studies there are on postpartum depression compared to other postpartum psychiatric illness, including obsessive–compulsive disorder, anxiety disorders, and postpartum psychosis. This lack spans the gamut of research techniques, from genetics to hormonal and immunological influences to imaging. This likely reflects at least partially the fact that postpartum depression is a common disorder and the other psychiatric disorders are far rarer. But, as the PACT consortium has shown, postpartum depression itself is a heterogeneous disorder with differing clinical phenotypes. This means that any two women diagnosed with postpartum depression could present with very different symptoms; it also means that there are symptom clusters that may cross the bounds of our clinical diagnoses and may prove better ways of understanding illness in the postpartum period. This enormous research gap, both in specific postpartum disorders other than depression and in symptom clusters that may cross diagnostic lines, represents an opportunity for future research as well as for comparisons among women with different clusters of symptoms.

Timing of onset of illness may matter

The postpartum time period carries with it an increased risk for the onset or relapse of psychiatric illness. Psychiatric illness may also begin during pregnancy and continue during the postpartum—either as new-onset psychiatric illness or as a recurrence from preexisting illness. It remains unclear if the biological processes that underlie psychiatric illness beginning during pregnancy are the same as those that underlie psychiatric illness beginning in the immediate postpartum time period. The importance of timing has been demonstrated several times in the current literature: for example, postpartum depression appears to have a familial or genetic basis, but only when onset is in the immediate postpartum (Forty et al., 2006; Murphy-Eberenz et al., 2006; Payne et al., 2008). Furthermore, two identified epigenetic biomarkers of postpartum depression exhibit different, indeed, opposite, methylation patterns depending on whether or not the woman was depressed during pregnancy (Guintivano et al., 2014; Osborne et al., 2015). Similarly, as reviewed in Chapter 6, obsessive–compulsive disorder may have very different presentations during pregnancy as compared to postpartum, with some studies showing that obsessions tend to focus on contamination fears during pregnancy and on harm to the child postpartum. Going forward, research into biological underpinnings should carefully note onset of illness and consider separating and comparing groups of women whose illness began in pregnancy from women whose illness began postpartum.

It remains unclear if perinatal psychiatric illness is biologically distinct from psychiatric illness outside of the peripartum

As research mounts, new-onset psychiatric illness during the peripartum time period is most often associated with psychiatric illness outside of the perinatal period—but not always. There is a subset of women who develop postpartum depression or postpartum psychosis, for instance, who never experience an episode of mood disturbance outside of the perinatal period. For example, among women who present with postpartum psychosis as their first psychiatric episode, 50%–80% will go on to have nonpuerperal episodes, mostly bipolar disorder (Chaudron & Pies 2003; Nager et al. 2013), meaning that 20%–50% do not. Because major depressive episodes often do not result in hospitalization (unlike postpartum psychosis and mania), the rate of women with first-onset postpartum depression who develop nonpuerperal episodes is hard to study—but clinically, a subset of women appear to have only postpartum depressive episodes and can discontinue medication use outside of the postpartum time. The rates of heritability of postpartum disorders are distinct from those outside the peripartum; as demonstrated in Chapter 1, the heritability of PPD is 44%–54%, while that of MDD outside the peripartum is only 32%. Is the biological basis of mood episodes that only occur in the postpartum time period the same as the basis for women who have both puerperal and nonpuerperal episodes? Studying this question will be difficult, as it may not be clear for years if a woman has only puerperal episodes.

Sensitivity to times of hormonal change may be present in a subset of women with psychiatric illness

A related research question is why do only some women with preexisting psychiatric illness become ill during the postpartum time period? The risk for psychiatric illness during pregnancy appears to be about the same as the risk outside of pregnancy (although many women with preexisting illness become ill during pregnancy due to stopping medications, often ill-advisedly). In the postpartum, however, the risk for mood and anxiety disorders increases dramatically—but not every woman with a psychiatric illness becomes ill at that time. Some of the factors that increase risk of psychiatric illness during the postpartum time period might include disturbed sleep, other environmental stressors, poor social support, personality vulnerabilities, or lack of pharmacological treatment of preexisting illness—yet the fact that not every women with a psychiatric illness becomes ill in the setting of the large hormonal changes that occur at delivery may imply that sensitivity to hormonal change may be a distinct biological factor that only some women have. This is supported by work done many years ago at the National Institute of Mental Health, in which women with a history of postpartum depression were shown to be "mood sensitive" to hormonal change while women without a history of postpartum depression were not (Bloch et al., 2000). A similar result was obtained in a sample of women with a history of premenstrual mood symptoms (Schmidt, Nieman, Danaceau, Adams, & Rubinow, 1998), as well as in a sample of women with a history of perimenopausal depression (Schmidt et al., 2015), suggesting that the sensitivity to hormonal change may be a distinct biological mechanism from the underlying biology for mood disorders in general. If so, this argues again for prospective studies that compare women with psychiatric illness and reproductive hormone sensitivity to women with psychiatric illness and no hormone sensitivity, as the underlying biology is at least partially different.

In most areas, it is premature to talk about biomarkers

While our group has identified and replicated epigenetic biomarkers that predict the development of PPD, such actual biomarkers are quite rare in perinatal psychiatric illness. In immune system research, for example, the immune changes of normal pregnancy are well characterized and some studies have shown differences in women who are psychiatrically ill—but these differences are not yet clearly defined and certainly cannot be crystallized into a single biomarker (despite the valiant attempts of some groups to create "inflammation scores" or other single measures, as described in Chapter 5). Similarly, while differences in allopregnanolone levels have been identified in both premenstrual dysphoric disorder and perinatal illness, we do not yet know if measuring these levels in individual women will be helpful, or if we need to look for other links in the chain of progesterone metabolism and/or GABA receptor signaling.

It is likely that a combination of factors, including genetics, hormonal changes, and immune system changes, underlie the pathogeneisis of postpartum psychiaric illness

It is becoming increasingly clear, at least in postpartum depression and postpartum psychosis, that there will not be one underlying biological cause but rather a number of biological variables that come together along with environmental factors to produce these illnesses. Furthermore, the illnesses that we call "postpartum depression" and "postpartum psychosis" are likely to be heterogeneous as well. Perhaps genetics plays a large role in some cases and not in others. Sensitivity to hormonal change may play a role in a subset of cases and environmental factors may play a larger role in others. Sorting out these different influences in individual cases will be difficult at best, but may be far easier in larger samples, particularly of women followed prospectively. Despite the complexity of identifying the biological basis for postpartum psychiatric illness, it is a worthy endeavor, and it is likely that sorting out the underlying biology will be easier in what may be a more homogenous group of patients than, for example, a random sample of patients with major depression or bipolar disorder.

Going forward it will be important to combine clinical observation with basic science approaches to identify biological causes of these illnesses. Postpartum onset may reflect a particular biological vulnerability that can be identified only when the sample of patients is more homogeneous and not diluted by individuals with onset during pregnancy or months after delivery. Identification of other clinical characteristics, such as anxiety or obsessions, that identify subgroups of patients that are more homogeneous may also be important. Imaging, genetic, immunological, and hormonal studies may also identify other important biomarkers that allow us to understand not only important subgroups of patients who are at risk but the bigger picture of how these illnesses develop and progress. Careful, thorough, and preferably prospective characterization of women at high risk of these disorders may eventually give us the data we need to identify psychiatric disorders, not by their clinical presentation as we presently do, but by their biological presentation. To complete such research, we will need to overcome the reluctance of many researchers and funding bodies to engage in rigorous research in a vulnerable population—an understandable reluctance, but one that is blocking the path forward for unlocking the causes of devastating illnesses that affect two generations.

References

Bloch, M., Schmidt, P. J., Danaceau, M., Murphy, J., Nieman, L., & Rubinow, D. R. (2000). Effects of gonadal steroids in women with a history of postpartum depression. *American Journal of Psychiatry, 157*(6), 924–930. https://doi.org/10.1176/appi.ajp.157.6.924.

Bratfos, O., & Haug, J. O. (1966). Puerperal mental disorders in manic-depressive females. *Acta Psychiatrica Scandinavica, 42*(3), 285–294.

Campbell, S. B., & Cohn, J. F. (1991). Prevalence and correlates of postpartum depression in first-time mothers. *Journal of Abnormal Psychology, 100*(4), 594–599.

Chaudron, L. H., & Pies, R. W (2003). The relationship between postpartum psychosis and bipolar disorder: A review. *Journal of Clinical Psychiatry, 64*(11), 1284–1292.

Cox, J. L., Holden, J. M., & Sagovsky, R. (1987). Detection of postnatal depression. Development of the 10-item Edinburgh postnatal depression scale. *The British Journal of Psychiatry, 150*, 782–786.

Davidson, J., & Robertson, E. (1985). A follow-up study of post partum illness, 1946–1978. *Acta Psychiatrica Scandinavica, 71*(5), 451–457.

Dietz, P. M., Williams, S. B., Callaghan, W. M., Bachman, D. J., Whitlock, E. P., & Hornbrook, M. C. (2007). Clinically identified maternal depression before, during, and after pregnancies ending in live births. *American Journal of Psychiatry, 164*(10), 1515–1520. https://doi.org/10.1176/appi.ajp.2007.06111893.

Forty, L., Jones, L., Macgregor, S., Caesar, S., Cooper, C., Hough, A., … Jones, I. (2006). Familiality of postpartum depression in unipolar disorder: Results of a family study. *American Journal of Psychiatry, 163*(9), 1549–1553. https://doi.org/10.1176/ajp.2006.163.9.1549.

Guintivano, J., Brown, T., Newcomer, A., Jones, M., Cox, O., Maher, B. S., … Kaminsky, Z. A. (2014). Identification and replication of a combined epigenetic and genetic biomarker predicting suicide and suicidal behaviors. *American Journal of Psychiatry, 171*(12), 1287–1296. https://doi.org/10.1176/appi.ajp.2014.14010008.

Kendell, R. E., Chalmers, J. C., & Platz, C. (1987). Epidemiology of puerperal psychoses. *The British Journal of Psychiatry, 150*, 662–673.

Murphy-Eberenz, K., Zandi, P. P., March, D., Crowe, R. R., Scheftner, W. A., Alexander, M., … Levinson, D. F. (2006). Is perinatal depression familial? *Journal of Affective Disorders, 90*(1), 49–55. https://doi.org/10.1016/j.jad.2005.10.006.

Nager, A., Szulkin, R., Johansson, S. E., Johansson, L. M., & Sundquist, K. (2013). High lifelong relapse rate of psychiatric disorders among women with postpartum psychosis. *Nordic Journal of Psychiatry, 67*(1), 53–58. https://doi.org/10.3109/08039488.2012.675590.

O'Hara, M. W. (2009). Postpartum depression: What we know. *Journal of Clinical Psychology, 65*(12), 1258–1269. https://doi.org/10.1002/jclp.20644.

Osborne, L., Clive, M., Kimmel, M., Gispen, F., Guintivano, J., Brown, T., ... Kaminsky, Z. (2015). Replication of epigenetic postpartum depression biomarkers and variation with hormone levels. *Neuropsychopharmacology*. https://doi.org/10.1038/npp.2015.333.

Payne, J. L., MacKinnon, D. F., Mondimore, F. M., McInnis, M. G., Schweizer, B., Zamoiski, R. B., ... Potash, J. B. (2008). Familial aggregation of postpartum mood symptoms in bipolar disorder pedigrees. *Bipolar Disorders, 10*(1), 38–44. https://doi.org/10.1111/j.1399-5618.2008.00455.x.

Payne, J. L., Roy, P. S., Murphy-Eberenz, K., Weismann, M. M., Swartz, K. L., McInnis, M. G., ... Potash, J. B. (2007). Reproductive cycle-associated mood symptoms in women with major depression and bipolar disorder. *Journal of Affective Disorders, 99*(1–3), 221–229. https://doi.org/10.1016/j.jad.2006.08.013.

Reich, T., & Winokur, G. (1970). Postpartum psychoses in patients with manic depressive disease. *The Journal of Nervous and Mental Disease, 151*(1), 60–68.

Russell, E. J., Fawcett, J. M., & Mazmanian, D. (2013). Risk of obsessive-compulsive disorder in pregnant and postpartum women: A meta-analysis. *Journal of Clinical Psychiatry, 74*(4), 377–385. https://doi.org/10.4088/JCP.12r07917.

Schmidt, P. J., Ben Dor, R., Martinez, P. E., Guerrieri, G. M., Harsh, V. L., Thompson, K., ... Rubinow, D. R. (2015). Effects of estradiol withdrawal on mood in women with past perimenopausal depression: A randomized clinical trial. *JAMA Psychiatry, 72*(7), 714–726. https://doi.org/10.1001/jamapsychiatry.2015.0111.

Schmidt, P. J., Nieman, L. K., Danaceau, M. A., Adams, L. F., & Rubinow, D. R. (1998). Differential behavioral effects of gonadal steroids in women with and in those without premenstrual syndrome. *New England Journal of Medicine, 338*(4), 209–216. https://doi.org/10.1056/nejm199801223380401.

Viguera, A. C., Tondo, L., Koukopoulos, A. E., Reginaldi, D., Lepri, B., & Baldessarini, R. J. (2011). Episodes of mood disorders in 2,252 pregnancies and postpartum periods. *American Journal of Psychiatry, 168*(11), 1179–1185. https://doi.org/10.1176/appi.ajp.2011.11010148.

Wesseloo, R., Kamperman, A. M., Munk-Olsen, T., Pop, V. J., Kushner, S. A., & Bergink, V. (2016). Risk of postpartum relapse in bipolar disorder and postpartum psychosis: A systematic review and meta-analysis. *American Journal of Psychiatry, 173*(2), 117–127. https://doi.org/10.1176/appi.ajp.2015.15010124.

Wisner, K. L., Perel, J. M., Peindl, K. S., Hanusa, B. H., Findling, R. L., & Rapport, D. (2001). Prevention of recurrent postpartum depression: A randomized clinical trial. *Journal of Clinical Psychiatry, 62*(2), 82–86.

Wisner, K. L., Perel, J. M., Peindl, K. S., Hanusa, B. H., Piontek, C. M., & Findling, R. L. (2004). Prevention of postpartum depression: A pilot randomized clinical trial. *American Journal of Psychiatry, 161*(7), 1290–1292.

Yonkers, K. A., Ramin, S. M., Rush, A. J., Navarrete, C. A., Carmody, T., March, D., ... Leveno, K. J. (2001). Onset and persistence of postpartum depression in an inner-city maternal health clinic system. *American Journal of Psychiatry, 158*(11), 1856–1863. https://doi.org/10.1176/appi.ajp.158.11.1856.

Index

Note: Page numbers followed by "f" indicate figures, "t" indicate tables and "b" indicate boxes.

R

S

Kolibri 360 GmbH
Ferdinand-Jühlke-Straße 7
99095 Erfurt, Deutschland
produktsicherheit@kolibri360.de